The Kimberley Craton:
The Purnululu National Park and the Bungle Bungle Range of Western Australia

By
William A. Szary, M.S.

Library of Congress Cataloging In Publication Data

Szary, William A.

Book Cover: An aerial view of the Bungle Bungle Range in the Purnululu National Park of the Kimberley Region of Western Australia. Source: Australia's Northwest posted on the internet.

Book Title: The Kimberley Craton:
The Purnululu National Park
and the
Bungle Bungle Range of Western Australia

Includes references.

ISBN-13: 9798436646688

Earth2Energy Educational Publishing
Port Richey FL 34668
Earth2Energy is a Registered Trademark

Chapter 1. Geology of the Kimberley Craton

Tyler and others (2012) presented a paper on the geologic evolution of the Kimberley region summarized in this chapter.

Overview

The history of the Kimberley region in the far north of Western Australia began in the Paleoproterozoic with rifting along the North Australian Craton margin at 1910–1880 Ma followed by plate collision as part of a series of 1870–1790 Ma events that formed the Diamantina Craton within the supercontinent Nuna. Collision involved the accretion of an intra-oceanic arc to a continent that included the Kimberley Craton before final collision and suturing with the North Australian Craton. The c. 1835 Ma Speewah Basin formed as a retro-arc foreland basin to the west. The post-orogenic c. 1800 Ma shallow-marine to fluvial Kimberley Basin and its equivalents had a provenance to the north and extended across both the Lamboo and Hooper provinces. Subsequent late Paleoproterozoic and Mesoproterozoic basins formed broadly similar depositional settings during break-up and reassembly into the supercontinent Rodinia. The intra-cratonic Yampi sinistral strike-slip faulting occurred between 1400–1000 Ma.

The Neoproterozoic Centralian Superbasin formed as a broad intra-cratonic sag basin throughout central Australia between c. 830 Ma and the earliest Cambrian including a series of basins across the Kimberley. Glacigene rocks are present with the most widespread being equivalent to the c. 610 Ma Elatina ("Marinoan") glaciation. Folding, thrusting, and strike-slip faulting during the c. 560 Ma King Leopold Orogeny caused a widespread unconformity at the base of the Ord and Bonaparte basins marked by the c. 508 Ma Kalkarindji Continental Flood Basalt Province. In the Early Ordovician, thermal subsidence initiated the Canning Basin. Paleozoic sedimentary rocks including Devonian reef complexes were deposited on the Lennard Shelf and in the Fitzroy Trough. In the Halls Creek Orogen, Devonian sedimentary rocks were deposited in sub basins of the Ord Basin during the c. 450–300 Ma Alice Springs Orogeny. A widespread glacigene succession followed in the Canning Basin, but by the early Triassic deposition was restricted and the remainder of the Mesozoic succession forms a veneer over much of the basin.

The geological history of the Kimberley region in the far north of Western Australia is complex and spans almost 2 billion years of earth history. A period of Paleoproterozoic plate collision during the assembly of the supercontinent Nuna produced crystalline basement rocks forming the Hooper and Lamboo provinces.

Collision and suturing was followed through the rest of the Proterozoic and the Phanerozoic by repeated phases of sedimentary basin formation, orogenic deformation, and associated fault reactivation at upper crustal levels, first within Nuna, then Rodinia, and finally Gondwana. Deposition essentially ceased in the Late Paleozoic, and in the Neogene the region began to bow downwards as Australia met the Indian plate (**Figures 1 and 2**).

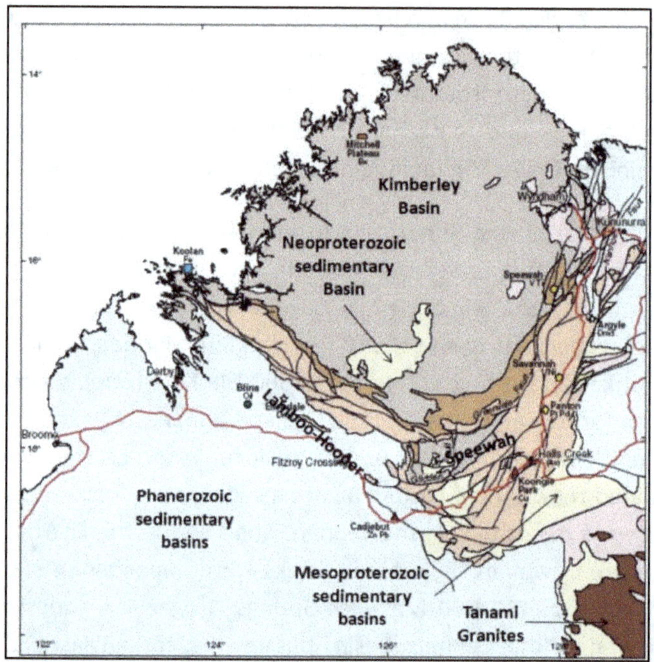

Figure 1. Tectonic units in the Kimberley region. Source: Tyler and others 2012.

Paleoproterozoic plate collision (1910–1805 Ma)

The oldest exposed rocks in the Kimberley region are igneous and low- to high-grade meta-igneous and meta-sedimentary rocks that occur within the 1910–1790 Ma Lamboo and Hooper provinces.

The Halls Creek Orogen is a suture formed by plate collision between the southeastern margin of a continent including the Kimberley Craton (which underlies the Speewah and Kimberley basins) and the NW margin of the North Australian Craton.

The Halls Creek Orogeny was one of a series of linked collisional events that formed the Diamantina Craton as part of the supercontinent Nuna. Collision was complex involving the accretion of the Tickalara arc to the Kimberley Craton during the 1870 –1850 Ma Hooper Orogeny before suturing during the 1835–1805 Ma Halls Creek Orogeny.

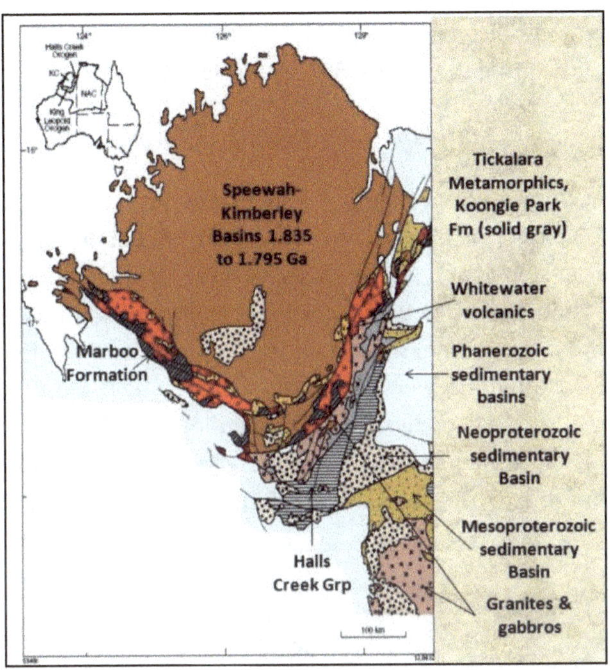

Figure 2. Paleoproterozoic geology of the Kimberley region. Source: Tyler and others 2012.

The Lamboo and Hooper Provinces

The Lamboo Province in the Halls Creek Orogen can be divided into three zones; the Western zone, the Central zone and the Eastern zone. The Hooper Province in the west Kimberley is an extension of the Western zone. Stratigraphic units cannot be correlated across the zone boundaries which is consistent with the zones forming Paleoproterozoic tectonostratigraphic terranes.

The oldest rocks in the Hooper Province and the Western zone of the Lamboo Province are low- to high-grade turbiditic metasedimentary rocks of the c. 1870 Ma Marboo Formation representing rifting marginal to the Kimberley Craton following accretion of earlier Paleoproterozoic exotic terranes. Partial melting of intermediate – felsic calc-alkaline rock within these source terranes formed the felsic volcanic rocks of the unconformably overlying the Whitewater Volcanics which were deformed and metamorphosed, and extensively intruded by potassic I-type granitic and sub-volcanic rocks as well as gabbroic rocks, and layered mafic–ultramafic intrusions of the Paperbark Supersuite during the 1865–1850 Ma Hooper Orogeny.

Central zone of the Lamboo Province

The Central zone is dominated by medium – high grade turbiditic metasedimentary and mafic volcanic and volcaniclastic rocks of the Tickalara Metamorphics interpreted as an oceanic island arc developed at c. 1865 Ma. These were intruded by tonalitic sheets, and deformed and metamorphosed between c. 1865–1856 Ma and at 1850–1845 Ma. In the southern part of the Central zone, sedimentary rocks and mafic and felsic volcanic rocks of the Koongie Park Formation were deposited at 1845–1840 Ma during rifting of the arc. Layered mafic–ultramafic bodies were intruded into the Central zone at c. 1856, c. 1845 and 1830 Ma. Large volumes of granite and gabbro of the Sally Downs Supersuite intruded the Central zone during the Halls Creek Orogeny at 1835–1805 Ma (**Figure 3**).

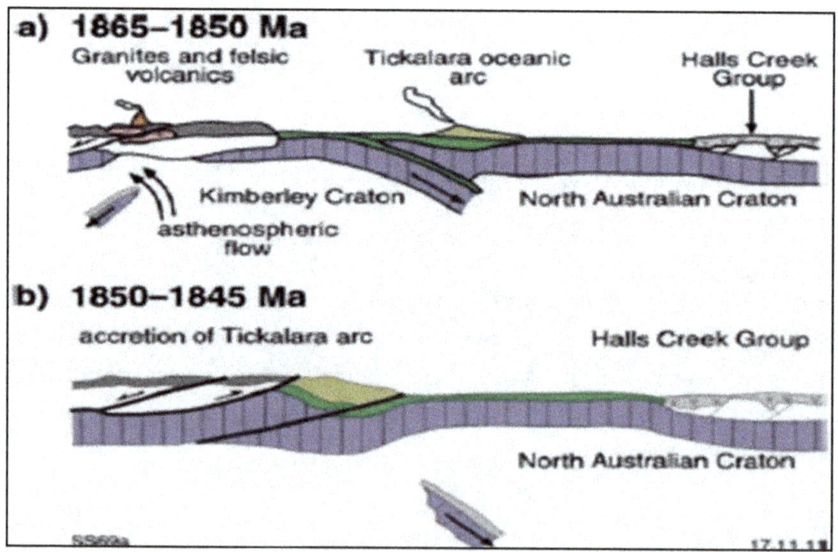

Figure 3. Plate tectonic setting during the Hooper Orogeny. Source: Tyler and others 2012.

The Eastern zone c. 1910 Ma mafic and felsic volcanic rocks of the Ding Dong Downs Volcanics and associated granitic rocks are unconformably overlain by low-grade metasedimentary and metavolcanic rocks of the Halls Creek Group. At the base of the Halls Creek Group, the quartz sandstone of the Saunders Creek Formation contains exclusively Archean detrital zircon populations (3600 Ma – 2512 Ma). Overlying mafic volcanic rocks of the Biscay Formation in the lower part of the Halls Creek Group were erupted at c. 1880 Ma on a passive continental margin along the western edge of the North Australian Craton.

The overlying turbiditic metasedimentary rocks of the Olympio Formation record a transition from a passive to an active margin and can be divided into upper and lower units separated by alkaline volcanism at c. 1857 Ma and c. 1848 Ma. The turbiditic rocks form part of two submarine-fan systems in an evolving foreland basin with sediment derived from predominantly continental granitic sources to the NW, the oldest at c. 1873 Ma and the youngest at c. 1847 Ma. The Halls Creek Group was deformed and metamorphosed during the 1835–1805 Ma Halls Creek Orogeny, and were stitched to the central zone by the 1820–1810 Ma granites of the Sally Downs Supersuite.

Hooper Orogeny

The Hooper Orogeny was recognized first in the Hooper Province taking place between c. 1870 Ma and c. 1850 Ma. Rocks of the Marboo Formation are affected by two early phases of deformation. The first was between c. 1870 Ma, the age of the youngest detrital zircons in the Marboo Formation, and c. 1865 Ma, the age of the intrusion of the oldest Paperbark Supersuite intrusions. The second deformation and accompanying metamorphism was coeval with the emplacement of the Paperbark Supersuite between 1865 and 1850 Ma, and deformed the c. 1855 Ma Whitewater Volcanics.

The first deformation in the Central zone post-dates the c. 1863 Ma age of the Rose Bore Granite while a minimum age is provided by the c. 1850 Ma granitic rocks of the Dougalls Suite which postdates the first deformation but pre-dates the second. These age constraints are similar to those for the second deformation in the Western zone suggesting both deformations are linked to the accretion of the Tickalara arc.

Halls Creek Orogeny

Deformation and metamorphism during the Halls Creek Orogeny affected the entire Lamboo Province. In the Central zone, the first deformation occurred synchronously with the intrusion of the c. 1835 Ma Mabel Downs Tonalite. The second deformation refolds earlier structures and is cut by c. 1810 Ma granite intrusions. The Sally Downs Supersuite was intruded between c. 1835 Ma and c. 1805 Ma and its granitic rocks range from syn-collisional crustally derived adakites through to post-collisional potassic types.

Speewah Basin

During the Halls Creek Orogeny, siliciclastic sedimentary rocks of the 1.5 km thick Speewah Group were being deposited in the c. 1835 Ma Speewah Basin unconformably overlying the northern and western margins of the Western zone of the Hooper and Lamboo provinces. The Speewah Group thins dramatically to the west, and is overlapped by the Kimberley Group in the SE.

The presence of c. 1834 Ma felsic volcanic rocks within the Valentine Siltstone near the base of the succession suggest the basin developed at the same time as the intrusion of granitic and gabbroic rocks into Lamboo Province at c. 1835 Ma. Paleocurrent data indicate a provenance from the NE from the core of the uplifted Lamboo Province with deposition taking place in a retro-arc foreland basin behind the active eastern margin of the Kimberley Craton.

Post-collisional Kimberley Basin and its equivalents (c. 1800 Ma)

The Kimberley Group comprises a 3 km-thick succession of siliciclastic sedimentary rocks and mafic volcanic rocks deposited in the Kimberley Basin which disconformably overlies the Speewah Group, but pre-dates the c. 1797 Ma Hart Dolerite. Paleocurrent data indicate a provenance from the north beyond the present Australian plate with deposition taking place in a broad semi-enclosed shallow marine basin. The Moola Bulla, Red Rock, Texas Downs and Revolver Creek basins are probable equivalents to the Kimberley Group that outcrop across the Halls Creek Orogen (**Figure 4**).

Hart Dolerite and Wotjulum Porphyry

The Hart Dolerite is a large igneous province that consists of a network of connected sills and dikes of massive dolerite and less extensive granophyres with a combined thickness of up to 3 km which are mainly intruded into the Speewah Group and lower Kimberley Group around their deformed SE and SW margins.

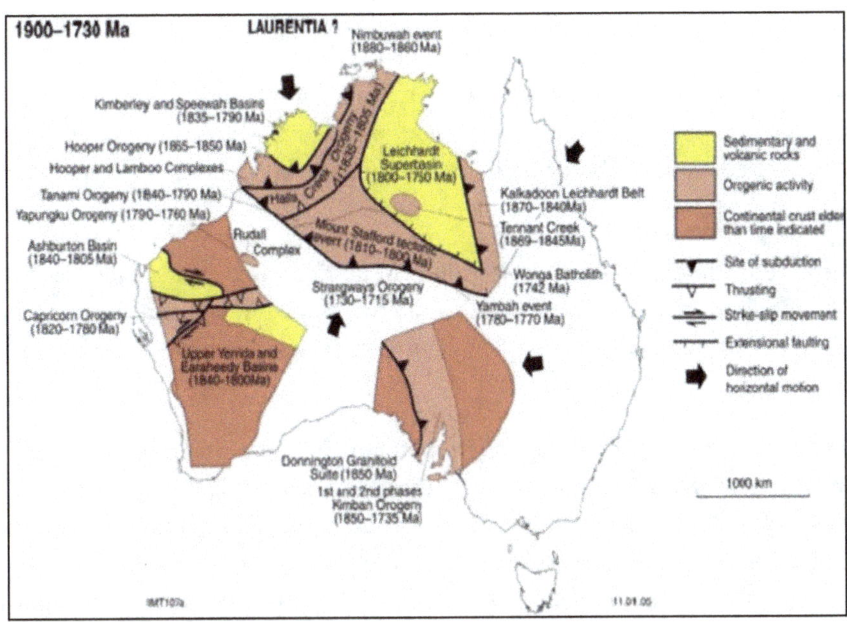

Figure 4. Model representing the assembly of the Diamantina Craton. Source: Tyler and others 2012.

The intrusions underlie an area of c. 160,000 km2 and have an estimated volume of 250,000 km3. The granophyres have been dated at c. 1797 Ma. The Hart Dolerite records part of a post-collisional magmatic event related to plate re-organization, and was sourced from subduction-modified mantle beneath the Kimberley Craton. The c. 1740 Ma Wotjulum Porphyry intrudes as a sill into the upper Kimberley Group in the Yampi Peninsula.

Intra-cratonic Proterozoic Basins and orogenies (1800–508 Ma)

Collision and suturing was followed by two broad cycles of sedimentary basin formation followed by intra-cratonic orogenic deformation and associated fault reactivation. The first spanned the late Paleoproterozoic and Mesoproterozoic as Nuna broke up and Proterozoic Australia was reassembled into Rodinia. The second reflects Neoproterozoic breakup of Rodinia followed by the assembly of Gondwana.

The ages of the Bastion, Crowhurst, and Osmond basins are uncertain but they all overlie the Kimberley Group or its equivalents regarded as younger than c. 1797 Ma. Based on limited isotopic dating and correlation with other units suggests deposition took place in the late Paleoproterozoic – early Mesoproterozoic.

Siliciclastic sedimentary rocks of the Bastion Group deposited in a similar environment to the Kimberley Group unconformably overlie the Kimberley Basin in the northeastern Halls Creek Orogen, and are overlain by the c. 508 Ma Cambrian Antrim Plateau Volcanics.

The Crowhurst Group which includes stromatolitic dolomite occupies a similar setting overlying the Kimberley Basin at the southern end of the Halls Creek Orogen, and the siliciclastic Mount Parker Formation and the overlying stromatolitic Bungle Bungle Dolomite were deposited in the Osmond Basin on the eastern side of the Halls Creek Orogen. The siliciclastic rocks of the 4.4 km-thick Carr Boyd Group unconformably overlie both crystalline rocks of the Lamboo Province and equivalents of the Kimberley Group in the northeastern Halls Creek Orogen. They are in turn unconformably overlain by Neoproterozoic glacial deposits of the Wolfe Creek Basin. Deposition of the Carr Boyd Group probably took place in a deltaic to shallow-marine setting at c. 1200 Ma, the age of intrusion of the Argyle lamproite diatreme into wet sediments. The Glidden Group and Wade Creek Sandstone have been correlated with the Carr Boyd Group.

Yampi Orogeny

The Mesoproterozoic Yampi Orogeny took place between c. 1400–1000 Ma based on K-Ar dating and Ar-Ar dating producing large-scale NE-facing folds and thrusts in the Speewah and Kimberley Basin successions on the Yampi Peninsula (Yampi Fold Belt).

Thrusts can be followed into NW-trending, SW-dipping ductile shear zones in the Hooper Province. Deformation was accompanied by a low- to medium-grade metamorphic event with the development of large porphyroblasts of garnet, andalusite, and staurolite. Large-scale strike-slip faulting took place in the Halls Creek Orogen during the Yampi Orogeny, and established a pattern of N-NE-trending synthetic sinistral faults, and E-NE-trending antithetic dextral faults consistent with the thrusting in the Yampi Fold Belt.

Neoproterozoic Basins

The Paleoproterozoic and Mesoproterozoic rocks are overlain unconformably by Neoproterozoic sedimentary rocks of the Wolfe Creek Basin, the upper Victoria River Basin, and the Louisa Basin in the Halls Creek Orogen, and by the Mount House Group and Oscar Range Group in the King Leopold Orogen. These linked basins are part of the Centralian Superbasin interpreted as a broad intra-cratonic sag basin that extended throughout much of central Australia between c. 830 Ma and the earliest Cambrian. Stromatolitic dolomite in the Ruby Plains group can be correlated with Supersequence 1. Several glacigene intervals are present with the most widespread being equivalent to the c. 610 Ma Elatina ("Marinoan") glaciation of Supersequence 3. Glacial striae are preserved across the Kimberley region.

King Leopold Orogeny

The c. 560 Ma King Leopold Orogeny produced extensive well exposed W-NW- trending fold and thrust structures in the King Leopold Ranges (Precipice Fold Belt) along the SW margin of the Kimberley Basin together with reactivation of shear zones in the Hooper Province. South-directed thrusting is linked to sinistral strike-slip faulting in the Halls Creek Orogen and deformation affected Neoproterozoic glacigene rocks. Deformation occurred about the same time as the Paterson Orogeny at the NE edge of the Pilbara Craton, and the Petermann Orogeny in central Australia.

Phanerozoic Basins

A thick Paleozoic and Mesozoic succession is present in the Canning Basin in the SW Kimberley, and in the Ord and Southern Bonaparte basins in the eastern and northern Halls Creek Orogen.

The King Leopold Orogeny gave rise to a widespread unconformity prior to extrusion of c. 508 Ma basaltic rocks of the Antrim Plateau Volcanics and deposition of middle – upper Cambrian sedimentary rocks in the Ord and Southern Bonaparte basins. The volcanic rocks are part of the extensive Kalkarindji Continental Flood Basalt Province that originally extended across 300,000 km2 of northern Australia.

In the Halls Creek Orogen, sinistral strike-slip faulting and associated folding and thrusting related to the c. 450 Ma – 300 Ma Alice Springs Orogeny in central Australia resulted in the development of a series of Devonian sub-basins (Ord Basin) adjoining basin bounding faults.

Cambrian rocks are absent SW of the Halls Creek Orogen. There, Ordovician and younger rocks of the Canning Basin unconformably overlie or are faulted against Proterozoic rocks. Ordovician rocks are exposed in a thin sliver along the northern margin of the basin on the Lennard Shelf beneath the exhumed Devonian reef complexes and associated conglomerates reflecting a much thicker Ordovician and Silurian succession in the subsurface of the central and southern Canning Basin.

Devonian reef complexes

Middle and Upper Devonian (Givetian, Frasnian, and Famennian) reef complexes form a belt of rugged limestone ranges some 350 km long and as much as 50 km wide on the Lennard Shelf along the northern margin of the Canning Basin. The reef complexes are spectacularly exposed as a series of fringing reefs, atolls, and banks which grew along the mountainous mainland shore of the Kimberley, and around rugged islands. Three main facies are recognized in the reef complexes: platform, marginal slope, and basin. The reefs are almost undeformed, with only local tilting due to normal fault movement (**Figure 5**).

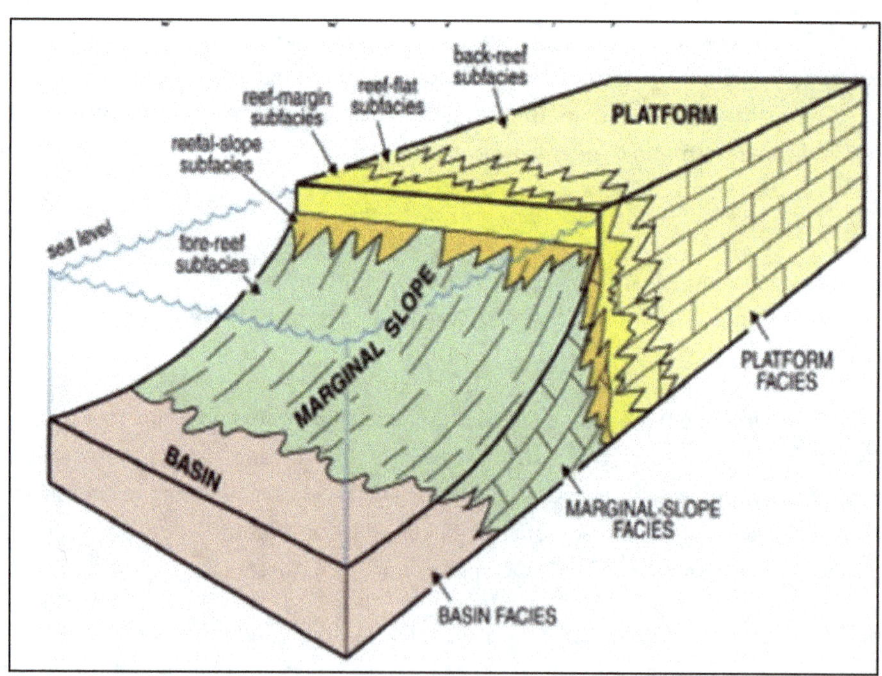

Figure 5. Reef complexes and facies subdivisions. Source: Tyler and others 2012.

Conglomerates interfinger with and cut through the reef complexes at prominent drainage notches related to the scarps of active faults in adjoining Proterozoic basement, and have been intersected (but not studied in detail) by some petroleum exploration drill holes basin ward of the outcrop belt. In outcrop, conglomerate bodies interfinger with cyclic platform carbonates. They span many meter scale cycles indicating deposition of each body extended over a few million years rather than tens or hundreds of thousands of years, suggesting tectonic rather than eustatic control.

The Famennian phase of conglomerate deposition is attributed to a major episode of the Alice Springs Orogeny of central Australia. The outcropping complexes preserve continued high stand deposits with corresponding low stand deposits preserved basin wards (to the S) in the subsurface. In outcrop, reefal development shows a broad transgressive-regressive cycle upon which several events are superimposed.

In the Givetian and Frasnian, reefs showed an overall back stepping pattern in which vertical reef growth was punctuated by abrupt back steps related to a sharp but small sea level fall followed by rapid transgression and drowning of the reef platform. These events appear to be irregular, and thus possibly of tectonic rather than eustatic origin. In the latest Frasnian at the peak of transgression and just prior to the Frasnian/Famennian global extinction event, there was rapid progradation and horizontal reef growth. Regression in the Famennian led to the reefs prograding in shallower water conditions, although subsidence still outstripped the rate of regression. Many Famennian reef builders migrated upwards from deeper water areas to take the place of newly extinct Frasnian builders such as stromatoporoids (**Figure 6**).

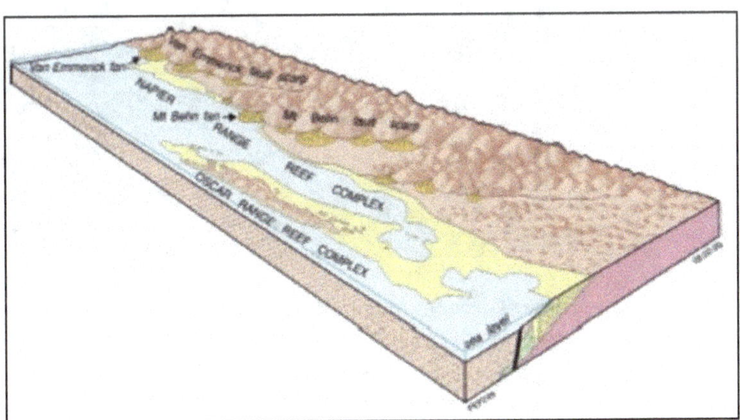

Figure 6. Block diagram based on actual geography of the Oscar and Napier Range reef complexes and associated conglomerates during the Frasnian time. Source: Tyler and others 2012.

In the Late Carboniferous and Early Permian continental-scale glaciation by north moving ice sheets possibly several kilometers thick planed off the tops of the limestone reefs. Beneath the ice sheets, subglacial water led to extensive karstification forming cave systems, karst corridors, solution dolines, and subglacial channels. Some of these channels are preserved today as the major gorges cutting through the reefs most notably Billy Munro Gorge, Windjana Gorge, and Geikie Gorge.

Clays of probable Permian age can be seen locally in some gorges and caves. Permian deposition was thickest and most continuous in the Fitzroy Trough south of the Lennard Shelf. There, deposition continued sporadically until the Late Permian in siliciclastic-dominated mixed deltaic to shelfal settings (**Figures 7, 8, & 9**).

Figure 7. Windjana Gorge exposes Carboniferous Pillara limestone in the wall on the left side of the Lennard River. On the right side of the river channel, a boulder of bleached limestone appears. Source: Windows 10 Spotlight posted on the internet.

A veneer of Mesozoic sedimentary rocks up to c. 200 m thick covers much of the central and southern Canning Basin thickening to 500–600 m near the coast. The Triassic succession characterized by a transgressive marine shale overlain by a prograding deltaic to fluvial succession is an onshore extension of a much thicker succession in the Roebuck Basin, and is largely restricted to the Fitzroy Trough and outer Lennard Shelf.

Figure 8. Geikie Gorge provides a clearer view of the Pillara limestone in contact with unbleached rock. Source: Budd Photography posted on the internet.

The Triassic succession is common to the Canning, Roebuck, Northern Carnarvon (where it hosts the giant gas fields of the NW Shelf) and Perth basins reflecting the onset of Gondwana breakup along Australia's NW margin. Along the seaward coast of the Dampier Peninsula, the Lower Cretaceous Broome Sandstone is notable for widespread dinosaur trackways preserved on coastal platforms as patchwork markings. The Broome Sandstone marks the start of a post-breakup transgression that extended over much of Australia's interior. In the Canning Basin, deposition was primarily of a fluvial to near shore, sandstone dominated veneer.

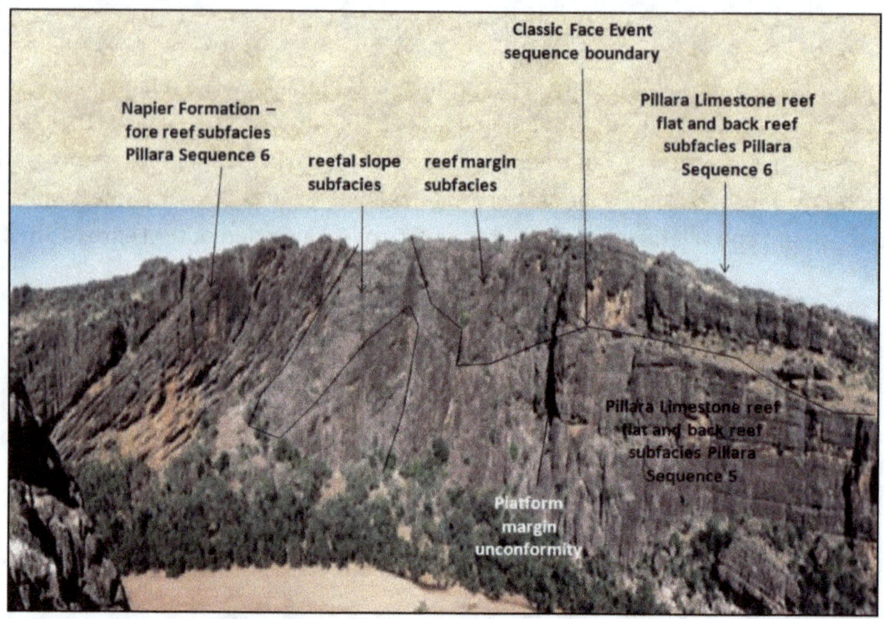

Figure 9. The Clastic Face at Windjara facing south showing transition from back reef deposits on the right through the reefal facies in the center, and the fore reef and marginal slope facies on the left. Source: Tyler and others 2012.

West and south of the Canning Basin (Northern Carnarvon and Gunbarrel basins), and over much of northern South Australia and western New South Wales, the transgression culminated in the Albian with deposition of a radiolarian-rich siltstone (**Figure 10**).

Evolving landscapes

The Kimberley region has an ancient landscape that has been evolving following the Neoproterozoic glaciations c. 600 Ma, Devonian tectonism in the Alice Springs Orogeny, Permian glaciation c. 280 Ma, and tilting of the Australian plate in the Neogene. At least two uplifted planation surfaces are present. The High Kimberley surface formed around 280–200 Ma and is preserved as remnants on the main Kimberley Plateau.

Figure 10. Aerial view of the Dampier Peninsular Ranges consisting of Broome Sandstone deposited by transgressive seas. Source: Roger's Website posted on the internet.

Between c. 200–100 Ma, uplift and erosion established the Low Kimberley surface surrounding the main plateau. Either or both surfaces may be related to rifting and Gondwana breakup offshore to the north and then northwest. Australia has drifted north towards India in the last 24 Myr with the Australian plate bowing beneath the Indian plate. Its southern margin has lifted up forming the various barrier plains of the Murray River in South Australia and the elevated Nullarbor Plain of the Eucla Basin. The northern continental margin has bowed downwards implying the northern Kimberley coastline may have been progressively drowning since the Miocene rather than simply during the Quaternary. Drowning has been accentuated by sea level rise after the last ice age c. 17 ka.

Mineral deposits and petroleum fields

The Kimberley region was the site of Western Australia's first Au rush in 1886 following the discovery of payable Au in the vicinity of Halls Creek in 1885. The small size of the deposits, the remoteness of the area, and the lack of fuel and water meant most of the prospectors left in 1892 on news of new discoveries near Kalgoorlie.

In 1951, hematitic Fe ore mines were developed by BHP on Cockatoo and Koolan islands in Yampi Sound a decade or more before the discovery and development of hematitic Fe ore in the Pilbara region.

Exploration for diamonds began in the late 1960's and witnessed the discovery of two major diamond provinces, Ellendale in the west Kimberley and Argyle in the east Kimberley. The Argyle diamond pipe (AK1), a world class deposit was discovered in 1979 with mining commencing in 1985. The area around Halls Creek remains highly prospective for Au and a number of small-scale alluvial and lode-gold mines have been developed. The Savannah Ni Mine north of Halls Creek together with the sub-economic Panton Point deposit highlights the considerable potential for the discovery of further mafic–ultramafic intrusion-related mineralization.

Zinc, Cu, and Pb occurs in volcanic-hosted massive sulfide deposits at Koongie Park SW of Halls Creek, and Mississippi Valley-type Zn-Pb deposits have been mined from the Devonian limestones of the west Kimberley at Pillara and Cadjebut. Bauxite deposits occur on the Mitchell Plateau, and represent deep weathering of underlying basalts in the Kimberley Basin. A small producing oilfield occurs at Blina in the Canning Basin, and huge reserves of gas are known in the offshore Browse Basin.

Chapter 2. The Purnululu National Park:

The Bungle Bungle Range

Hoatson and others (1997) presented a publication on the Bungle Bungles Range in Purnalulu National Park summarized in this chapter.

Geologic Setting

The Bungle Bungle Range of the Purnululu National Park in the remote East Kimberley of northwestern Australia is renowned for its spectacular beehive-shaped sandstone towers and gorges. The range is a deeply dissected plateau formed of sandstone and conglomerate representing sand and gravel deposited during the Devonian period about 360 million years ago. However, the beehives and gorges are much younger than this being a result of erosion that has taken place by flowing water in creeks and rivers during the last 20 million years.

Extending for 25 km from north to south and 30 km from east to west rising to about 250 m above the surrounding sand plain, the Bungle Bungle Range is an intricate maze of beehive-shaped peaks, narrow gorges lined with majestic *Livistona* fan palms, and soaring cliffs with many seasonal waterfalls and rock pools. Major attractions are Echidna Chasm, Froghole and Mini Palms gorges in the northwest, and Piccaninny and Cathedral gorges in the southwest.

General Information

The Bungle Bungle Range and surrounding area have supported a rich Aboriginal culture for at least 20,000 years, but it was little more than a century ago in 1879 that members of Alexander Forrest's exploration party became the first white Australians to see the Bungle Bungle Range. Part of the mystique of the Purnululu region is that until recently it was known only to the local Aboriginal people, pastoralists, stockmen, and a few others. Only after widespread media promotion in 1983 was the brilliance of this 'hidden jewel' of the East Kimberley uncovered. In recognition of the area's unique landscape, significance to the Aboriginal people, botanical importance, and tourist potential. The Western Australian Government established the Purnululu National Park (239,723 hectares) and the adjacent Purnululu Conservation Reserve (79,602 hectares) on the 27th of March 1987. Purnululu National Park is now regarded as one of Australia's premier scenic attractions, rivaling other famous symbols of outback Australia such as Uluru (Ayers Rock) and Kata Tjuta (The Olgas) (**Figure 11**).

Figure 11. Satellite view of the Bungle Bungle Ranges in Purnululu National Park.

Purnululu National Park is open to drive-in visitors from 1 April to 31 December (weather conditions permitting) with the most popular period being between May and September. The climate of the Kimberley region is classified as dry monsoonal and is characterized by hot wet summers lasting from November to March (the wet season) and warm dry winters from April to October (the dry season). The actual start and finish of these two contrasting seasons vary from year to year.

Average daily maximum temperatures during the dry season range from 29°C in July to 38°C in October. Little rain falls during this period and the humidity is relatively low. Temperatures may remain high at night, but frosts can occur at dawn during the cooler months of June, July, and August. From October onwards the temperature rises, humidity increases, and thunderstorms become common making conditions less comfortable for most visitors. During the wet season, the daily maximum temperature hovers around 40°C. Most of the annual rainfall of 500 to 700 mm occurs between 1 December and March. Heavy rains flood creeks turning soils to mud that make tracks impassable.

Vehicle access to Purnululu National Park and the Bungle Bungle Range is via the unsealed Spring Creek Track which leaves the Great Northern Highway about 250 km (by road) south of Kununurra 110 km northeast of Halls Creek. A shelter with information boards has been erected by CALM at the side of this track near the turn-off.

The Spring Creek Track is restricted to four-wheel-drive vehicles as it has many rocky and narrow sections with sharp bends, steep grades, and deep creek crossings. Caravans are prohibited. All vehicles are required to engage four-wheel-drive from the highway turn-off to help reduce damage to the track. The drive of 53 km to the ranger station within the national park generally takes about 2.5 hours so that travelling times from Kununurra and Halls Creek to the ranger station are about 5.5 and 4 hours, respectively.

Park entry fees are currently paid at a self-registration bay near the ranger station. Souvenir items (shirts, booklets, postcards, etc.) can be purchased from the shop on the verandah of the ranger station, but there is no provision for buying food. A short distance southeast of the ranger station, the Spring Creek Track meets the Gorge Track at Three-Ways Junction. The Gorge Track runs along the western side of the Bungle Bungle Range and provides access to the campsites and walking trails in the Purnululu National Park. A field tour is offered in the following chapter.

Commercial operators currently conduct full- day and longer safari drive tours into the park. Camping is restricted to three designated areas on the western side of the Bungle Bungle Range. These are Kurrajong (7 km north of the ranger station), Walardi (12 km south of the ranger station), and Bellburn (13 km south of the ranger station). The Bellburn area is for use only by licensed operators of fly/drive tours. Kurrajong and Walardi cater accommodates individual and group camping. Both have quiet sites and areas where generators can be used (**Figure 12**).

Geological Formations

The beehives and gorges of the Bungle Bungle Range in Purnululu National Park are carved from sandstone and conglomerate which were deposited as sand and gravel in a large shallow depression known as the Ord Basin deposited during the Devonian period (about 360 million years ago). The sandstone and conglomerate are encircled by older rocks of Cambrian age (550 to 500 million years) which include the limestone that forms a prominent wall on the plain west of the Bungle Bungle Range. The oldest Cambrian rocks are dark basalt lava flows forming a series of low ridges west of the limestone wall along the western edge of the national park.

Much older rocks ranging in age from around 600 million years to 1880 million years or even older form the hills and ridges farther west between the Purnululu National Park and the Great Northern Highway. They include examples of the three major classes of rocks: sedimentary, igneous, and metamorphic. In general, the rocks in the Purnululu National Park region increase in age from the Bungle Bungle Range westwards towards the Great Northern Highway.

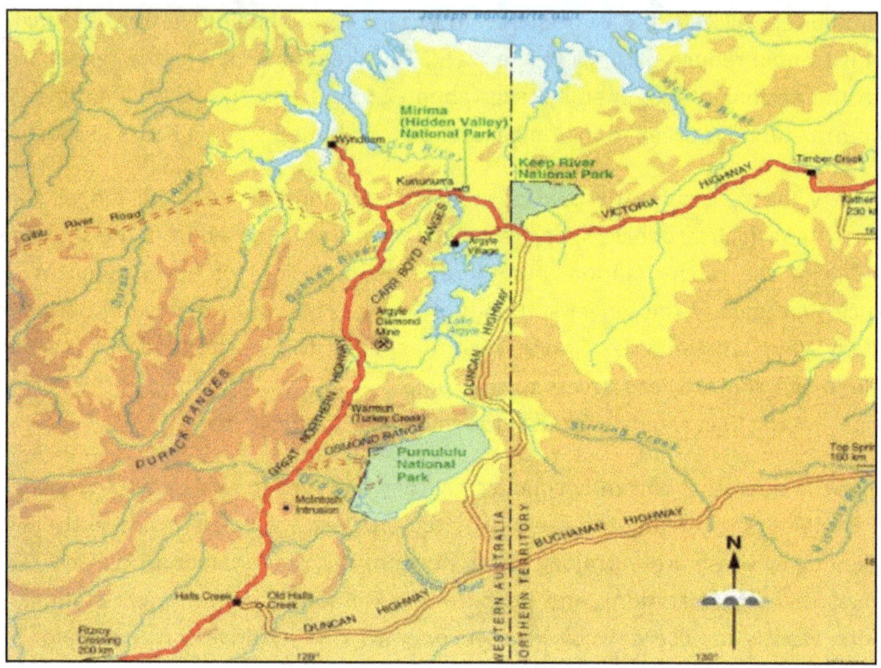

Figure 12. General location map of the Purnululu National Park. Source: Tyler and others 2012.

The Halls Creek Fault, a major fracture in the Earth's crust, passes through Calico Springs about midway between the national park and the Great Northern Highway. This fault is marked by a high ridge running north-south separating igneous and strongly metamorphosed rocks to the west from weakly metamorphosed sedimentary and igneous rocks to the east. The geological map shows the distribution of the main rock units in the area (**Figure 13**).

Geologic Age

A commonly asked question about any rock formation is 'How old is it?' In eastern Australia the rock itself may be a few hundred thousand to hundreds of millions of years old. In central Australia many of the rocks are between one and two thousand million years in age.

In the Yilgarn and Pilbara regions of Western Australia some rocks are as old as 3700 million years. The oldest rocks known in the Kimberley, about 1900 million years old occur to the south of Purnululu National Park. To put such ages in perspective, the Earth was formed approximately 4560 million years ago and the oldest rocks known on Earth which are in Canada are dated at about 4000 million years.

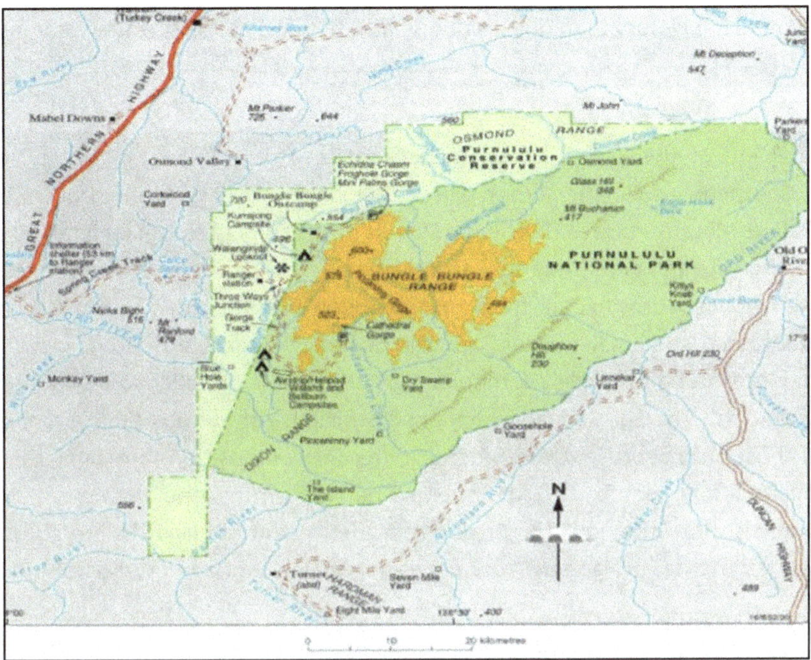

Figure 13. Purnululu National Park map showing significant features. Source: Tyler and others 2012.

If a calendar year is equated to the Earth's history of 4560 million years, a single day corresponds to 12.5 million years, one hour to 520,550 years, and one second to 144.6 years. On this 'calendar clock', blue-green algae (known as cyanobacteria), the earliest evidence of life on Earth, appeared in late March (about 3500 million years ago).

A wide variety of life forms including jellyfish, worms, trilobites, and animals with backbones burst onto the scene around 19 November (545 million years ago). The sand and gravel represented by the sandstone and conglomerate of the Bungle Bungle Range were deposited by rivers on 3 December (360 million years ago).

Dinosaurs first roamed the Earth on 14 December (225 million years ago) and became extinct on 27 December (65 million years ago). The sandstone beehives and gorges of the Bungle Bungle Range started to form by erosion at 10.00 am on 30 December (20 million years ago). 'Man' (*Homo sapiens*) did not appear on Earth until about 11.25 pm on 31 December (300,000 years ago), and Aborigines were living in the Purnululu region at 11.57 pm on 31 December (20,000 years ago). Alexander Forrest 'discovered' the Bungle Bungle Range less than one second before the end of the calendar year in 1879.

The Spring Creek Track

The first 9 km of the Spring Creek Track from the turn-off on the Great Northern Highway southeast to Fletcher Creek crosses low hills formed of Mabel Downs Tonalite and Tickalara Metamorphics. The Mabel Downs Tonalite (a granite-like igneous rock) is made up of crystals of glassy quartz, pale feldspar, and dark biotite and hornblende that crystallized slowly from magma (molten rock) deep in the Earth's crust. But before it crystallized about 1830 million years ago, the magma intruded (pushed its way) into the Tickalara Metamorphics. The dark minerals in the tonalite are aligned giving the rock a strong 'foliation'. The tonalite also contains 'inclusions' of various finer grained rocks. Some of these may represent a magma of different composition, but similar in age to the tonalite, and some are fragments derived from the adjacent Tickalara Metamorphics. Both the igneous tonalite and the metamorphic rocks are cut by veins filled with pegmatite, an igneous rock consisting mainly of large crystals of quartz and feldspar.

The Tickalara Metamorphics along this part of the Spring Creek Track represent sedimentary and igneous rocks that became involved in a 'mountain-building' event more than 1830 million years ago. During this event, the rocks now seen on the surface were buried at depths of many kilometers, and were crumpled and buckled as if squeezed in a giant vice. At the same time, they were subjected to extreme pressures and temperatures to the extent that in places they began to melt. As a result, the sedimentary and igneous rocks were transformed (metamorphosed) into high-grade crystalline metamorphic rocks (**Figures 14 & 15**).

Figure 14. Aerial view of Purnululu National Park. Figure 15 is a geologic map of the same region.

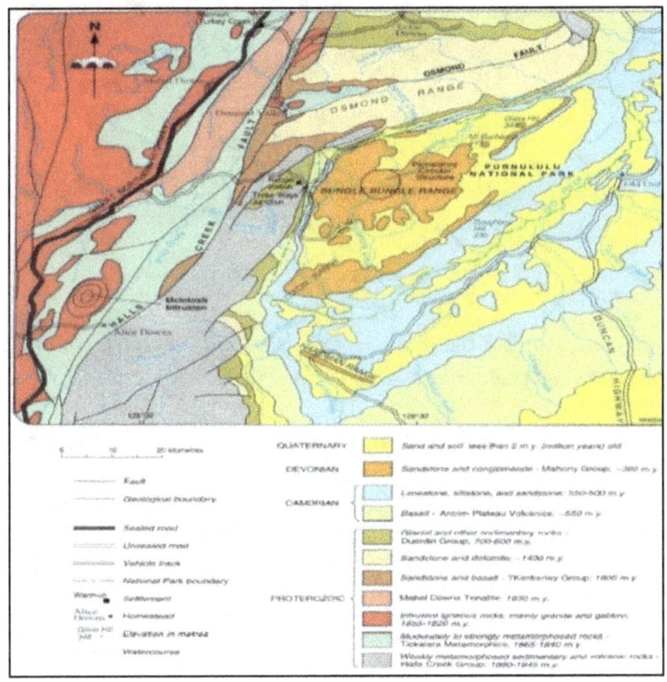

Figure 15. Geologic map of the Purnululu National Park. Source: Tyler and others 2012.

Stop 1. Fletcher Creek and the Ticklalara Metamorphics

Crystalline rocks of the Tickalara Metamorphics can be examined on both sides of the track in the sandy bed of Fletcher Creek. These are complex rocks consisting of dark-colored material rich in biotite and hornblende, and light-colored material of mainly quartz and feldspar some of which forms cross-cutting veins. After crossing Fletcher Creek, the track traverses more Mabel Downs Tonalite before turning northeast to stop 2 following the contact between the tonalite to the left of the track and Tickalara metamorphic rocks to the right of the track. The metamorphic rocks here include ridge forming bands of marble representing originally flat-lying but now folded beds of limestone (**Figures 16, 17, & 18**).

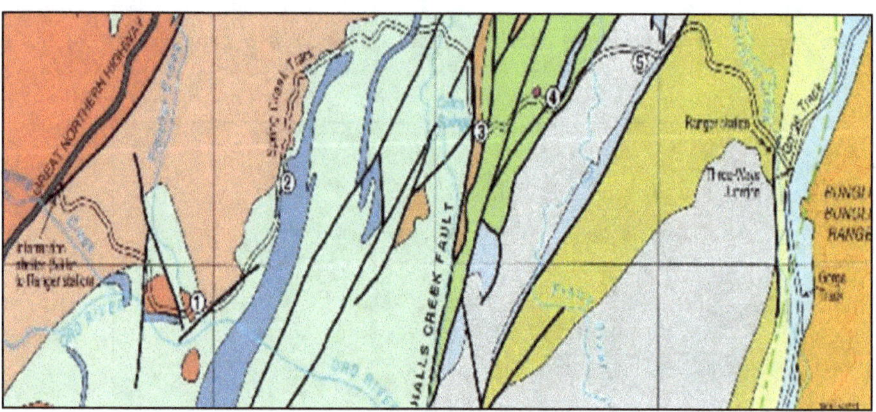

Figure 16. Geologic map showing the first five stops through the national park. Refer to Figure 14 for geologic formation units. Source: Tyler and others 2012.

Stop 2. Tickalara Metamorphics and Mable Downs Tonalite

Tight complex folds in marble of the Tickalara Metamorphics can be seen at this stop on the right-hand (eastern) side of the track 16 km from the highway. Banding in the marble represents sedimentary layering in the original limestone. Calcium carbonate derived from the marble has cemented some of the soils here to produce a hard crust known as calcrete.

The track continues north and then northeast mainly across Mabel Downs Tonalite and Tickalara Metamorphics, before turning south towards the permanent waterholes at Calico Springs.

Figure 17. Composite aerial of the Spring Creek Track showing locations of Stops 1 through 5. The west end marks the Great Northern Highway entrance opposite the Spring Creek Campground. The east end marks the Three Ways Junction.

Figure 18. The sandy bed at Fletcher Creek (Stop 1) exposes Tickalara Metamorphic rock on both sides of the track crossing.

Calico Springs lies on the Halls Creek Fault at a site where several major sub-parallel vertical faults come together. The high ridges here are formed of sandstone and conglomerate (? Kimberley Group) which were deposited as flat-lying beds about 1800 million years ago and later were steeply tilted during movements along the faults. The Halls Creek Fault can be traced for over 850 km south-southwest from near Darwin in the Northern Territory to the edge of the Great Sandy Desert south of Halls Creek. Matching geological features across the Halls Creek Fault indicate that the rocks on the eastern side have probably moved many kilometers to the northeast relative to those on the western side. This type of sideways movement with only a little vertical displacement is similar to that along the destructive San Andreas Fault in California (**Figure 19**).

Figure 19. The Calico Springs is located in the green drainage area to the right of the large ridge where the track curves to the northeast. The Hall Creek Fault is beneath the track segment oriented to the northeast up to the curve to the east. The small ridge at the curve to the east is a synclinal structure. On the south side of the track, a small ridge is offset to the left of the main ridge north of the track at the curve to the east point supporting interpretation of the fault trace.

Several evidences are present in Figure 19 of faulting occurring at Stations 1, 2, and 3. At Station 1, the track along the south end of the prominent ridge coincides with a strike slip fault evidenced by the offset of the lower ridge on the east side of the green vegetated drainage channel. The fault is a dextral slip fault. At Station 2 in the vegetated channel, the Calico Spring is evidence that ground water is leaking from faulting activity along the Halls Creek Fault. At Station 3, the axis of a synclinal channel between the tan tilted beds on the west and the juxtaposed layered sequenced beds on the right suggest faulting may be present although the strike slip fault along the track may have pushed the tan western ridge up against the eastern bedded sequence.

The track continues eastwards across rolling hills and plains formed on ancient basalt lava flows. These flows were erupted from volcanic vents shortly after the deposition of the conglomerate and sandstone at Calico Springs.

A small hill on the left-hand (northern) side of the track consists of much younger peridotite (an igneous rock made up mainly of the mineral olivine) that fills a pipe-like body cutting through the basalt. Within this general area of basalt there are also several steep-sided black ridges of older calcareous sedimentary rocks that belong to the 1850 million year old Halls Creek Group. Tight folds visible in these older rocks were formed long before the basalt lavas were erupted.

Stop 4. Steep Side Black Ridges

The track here passes just to the right of one of the steep-sided black ridges. Fallen blocks of thinly banded white, grey, and pink calcareous sedimentary rocks and subordinate red jasper showing small-scale folding lie near the base of the ridge. A large and much tighter fold leaning steeply to the west can be seen on the upper part of the vertical rock face above.

From stop 4 the track continues northeast across ancient basalt lavas to the Frank River crossing where a major fault separates the basalt from an older folded sequence of weakly metamorphosed mudstone and sandstone to the east. These older sedimentary rocks which belong to the Halls Creek Group like the calcareous rocks forming the steep-sided black ridges to the west give rise to a complex pattern of narrow steep-sided low ridges. The track keeps mainly to the crests of the ridges, and hence follows a highly sinuous course with many sharp bends.

Folds in metamorphosed mudstone and sandstone can be seen in a small cutting on the right-hand (southern) side of the track at this stop. Much larger spectacular folds in the same rocks to the south are prominent from the air (**Figure 20**).

Continuing eastwards, the track soon crosses another major fault marked by a steep-sided sandstone ridge which separates the folded rocks to the west from a much younger sequence that was not metamorphosed of mudstone and muddy sandstone (Ranford Formation). These rocks were deposited about 650 million years ago, continuing eastwards along the track as far as the ranger station and information booth. At the base of this rock sequence there is a unit of tillite, a distinctive rock type formed of rock debris and scattered boulders ('erratics') deposited by ancient glaciers. Many of these erratics show scratches and scours formed as they rubbed against other rocks while carried along by the ice. Tillite of similar age is known in other parts of the Kimberley and also in central and southern Australia indicating that much of the continent was covered by ice at this time. With increasing global temperature, the ice sheets melted and deposited tillite. The tillite can be seen resting on the older folded rocks of the Halls Creek Group a few hundred meters to the west of the ranger station.

Figure 20. Low hills adjacent to the track at Stop 5 consist of folded sandstone and mudstone.

From the ranger station southeast to Three- Ways Junction, the track crosses a younger unit of basalt lavas called the Antrim Plateau Volcanics. These lavas which encircle the Purnululu National Park were erupted onto the mudstone and sandstone of the Ranford Formation at the beginning of the Cambrian period about 550 million years ago.

A few kilometers east of Three-Ways Junction, the western escarpment of the Bungle Bungle Range rises 250 m above a plain on which there are several low ridges of Cambrian sedimentary rocks. The sandstone beehives for which the range is famous cannot be seen from this location as they are largely confined to the southern and eastern parts of the range.

The Gorge Track

The Spring Creek Track from the west meets the north-south-trending Gorge Track at Three-Ways Junction. The Gorge Track provides access to the northern and southern gorges and the campsites of the Purnululu National Park. It mainly crosses Cambrian sedimentary rocks largely concealed by sand which overlie the basalt lavas of the Antrim Plateau Volcanics. The Cambrian rocks underlie the Devonian sandstone and conglomerate that form the Bungle Bungle Range. They include a thin band of grey limestone (Headleys Limestone) forming a low but prominent wall like ridge along part of the track, and an inconspicuous band of younger limestone and shale (Linnekar Limestone).

The limestones were deposited as calcareous mud and silt in a shallow warm sea about 525 million years ago. In the southern part of the Purnululu National Park, the Linnekar Limestone contains fossils of trilobites (mostly *Redlichia),* small conical shells *(Biconulites hardmani),* and algae. The youngest Cambrian rocks visible from the Gorge Track are dark reddish sandstones about 520 million years old which form a relatively low ridge north of Three-Ways Junction to the right (east) of the track, and are also present along the basal part of the western escarpment of the Bungle Bungle Range to the south. The red sandstones contain desiccation cracks and trilobite tracks indicating that they were deposited on intertidal flats (**Figure 21**).

Northwards from Three-Ways Junction, the Gorge Track passes turn-offs to Walanginjdji Lookout (about 1.5 km from Three-Ways Junction), Kurrajong campsite (6 km), and Froghole Gorge carpark (18 km) before finishing at the Echidna Chasm carpark (19 km). For the first few kilometers north from Three-Ways Junction, it crosses basalt lavas (Antrim Plateau Volcanics) before going through a gap in a narrow wall of steeply tilted limestone (Headleys Limestone).

The track continues northwards across a sand plain between the limestone wall to the left (west) and a ridge of younger Cambrian red sandstone to the right (east), and then swings northeastwards towards Echidna Chasm with the imposing cliffs of the Bungle Bungle Range to the right (southeast) (**Figures 22, 23, & 24**).

Figure 21. Traveling north from the Three Way Junction at the first creek crossing, small ridges of gray limestone belonging to the Linnekar Limestone formation are exposed in the hills on the north side of the Gorge Track.

Figure 22. From the Froghole Gorge carpark on the north side of the Bungle Bungles, a southeast trail leads to the western escarpment of the Bungle Bungle Range.

Figure 23. A second trail leads to east of north towards the Echinda Chasm. Black cyanobacteria covering coats the tilted sandstone exposed in the mid-ground view. See Figure 24 for detailed photograph.

Figure 24, The large tilted dark rock seen in the background on the trail leading to the Echidna Chasm belongs to a set of sandstones viewed up close. The dark coating on the rock is due to cyanobacteria.

South from Three-Ways Junction, the Gorge Track follows the wall of steeply tilted grey limestone. Good views of the western escarpment of the Bungle Bungle Range can be obtained from the top of this limestone wall. At a T-intersection about 10 km south of Three-Ways Junction, the Gorge Track turns sharply left (east) for Piccaninny Creek whereas the other branch of the track continues south for about 2 km to the Walardi and Bellburn campsites. Proceeding eastwards across soft sand, the Gorge Track passes a turn-off to the right that leads to Purnululu airstrip and helipad 15 km from Three-Ways Junction, and climbs to the top of a broad sand dune. From here the track turns north towards the southern side of the Bungle Bungle Range marked by tiers of beehives (**Figure 25**).

On the way the track passes close to several isolated groups of beehives including those of Elephant Rock on the right-hand (eastern) side of the track. After crossing a rocky creek bed, the track eventually ends at the Picaninny Gorge carpark about 24 km from Three-Ways Junction. This carpark is the starting point for the Picaninny Gorge Trail, Cathedral Gorge Trail, and Domes Trail (**Figure 26**).

Figure 25. The Bungle Bungle Range exposed from the track leading into the southern side of the national park attraction.

Figure 26. The rocky creek bed appears to be Devonian Malony Sandstone lying beneath the sand plain on the approach to the Picaninny carpark.

The Bungle Bungle Range is an erosional remnant of an extensive blanket of sand and gravel that filled the Ord Basin about 360 million years ago during the Devonian period. Most of the sand was deposited by braided rivers flowing mainly from the northeast, but some formed sand dunes built up by winds from the south and southwest. At the same time, thick gravels in the northwest were deposited as alluvial fans that flanked mountains farther to the northwest (these mountains have long since been eroded away). The sand and gravel were progressively buried by younger sediments (no longer present), and became compacted to form sandstone (Glass Hill Sandstone and Purralili Sandstone) and conglomerate (Boll Conglomerate). These sedimentary rocks make up the Mahony Group (**Figure 27**).

The Boll Conglomerate in the northwestern end of the range consists of pebbles, cobbles, and boulders of quartzite, sandstone, and rare igneous rocks embedded in a sandy matrix. Most of the pebbles, cobbles, and boulders are smoothly rounded indicating that they were tumbled along river beds for considerable distances before being incorporated in the alluvial fan deposits. Some boulders are reported to show glacial scratchings, and may have come from the erosion of tillite like that found near the ranger station.

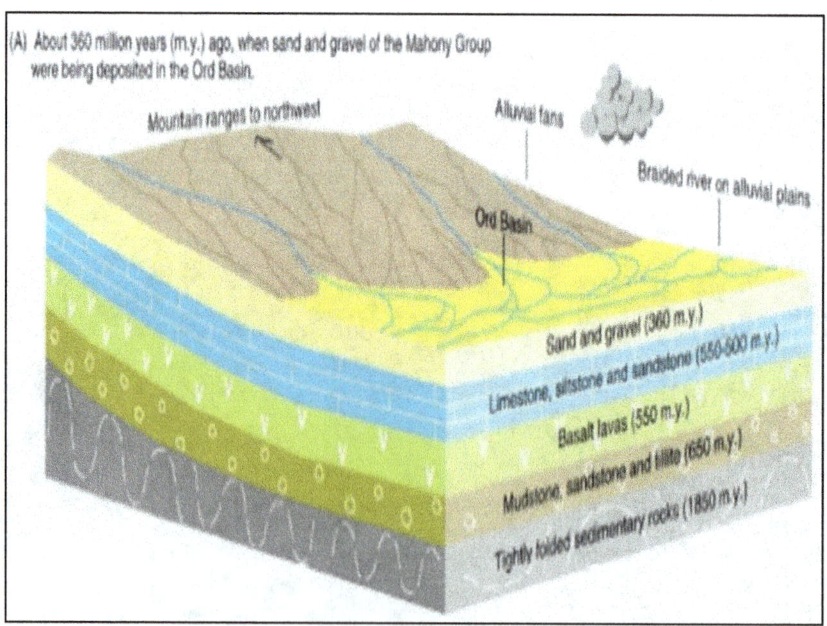

Figure 27. The Bungle Bungles began to develop in the Ord Basin about 360 mya. Source: Tyler and others 2012.

The rocks of the Bungle Bungle Range do not contain any well-preserved fossils which can be used for age dating. Their Devonian age is based on a correlation with similar sandstones containing fossils known to be Devonian near Kununurra to the north and near Billiluna homestead to the south of Halls Creek.

At the same time as the Devonian fluvial sands and gravels of the Bungle Bungle Range were being deposited on land in the East Kimberley, limestone reefs were growing in a shallow sea covering part of the West Kimberley. These limestone reefs form the present-day Napier Range, and are well-displayed in the Windjana Gorge, Tunnel Creek, and Geike Gorge national parks.

Some 300 million years ago, after the sandstone and conglomerate of the Bungle Bungle Range were buried to a depth of several kilometers by younger sediments, the East Kimberley was involved in an orogeny — a period of mountain building. The main effects of this orogeny evident in the Purnululu National Park were the uplifting and tilting of the Devonian and underlying rocks in the west and northwest. During this orogeny, movements took place along many of the major faults in the region including the Halls Creek Fault which passes through Calico Springs.

The summits of the Bungle Bungle Range and surrounding high ridges are remnants of an old nearly flat land surface generally lying 500 to 600 m above the present-day sea-level and more than 200 m above the Ord River plain. This old land surface which can be traced over much of the East Kimberley and into the Northern Territory resulted from prolonged erosion that followed the orogeny 300 million years ago. Several kilometers thickness of rock was removed as erosion continued until about 20 million years ago when the region had been reduced to a low-level undulating plain resembling that of the present-day Tanami Desert in central Australia (**Figures 28 & 29**).

The last 20 million years, though has seen a dramatic lowering of the sea-level, or uplift of the land in this part of northern Australia. As a consequence of the sea-level change, the Ord River and its tributaries have cut down into the old land surface carving out the beehives, gorges, and cliffs of the Bungle Bungle Range.

Jointing in the sandstone and conglomerate of the Bungle Bungle Range has played an important role in the formation of the gorges and beehives. Joints are natural planes of weakness or fractures that typically form during release of pressure as overlying rocks are removed by erosion. They commonly have a regular pattern orientated both perpendicular and parallel to the land surface. Water flowing across rock surfaces becomes channeled along the joints.

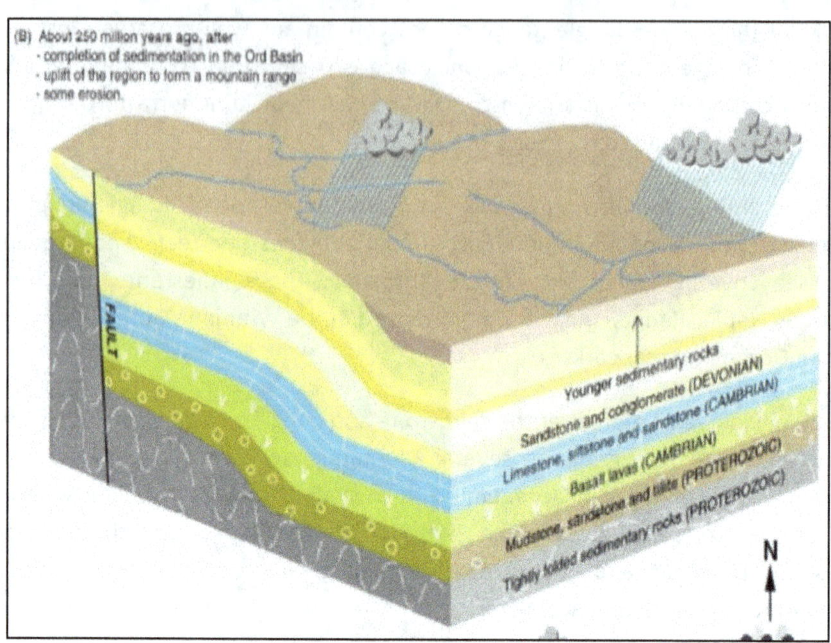

Figure 28. About 250 mya, sedimentation was completed in the Ord Basin. Uplift occurred along faults during mountain building. Some erosion began during uplift and deformation. Source: Tyler and others 2012.

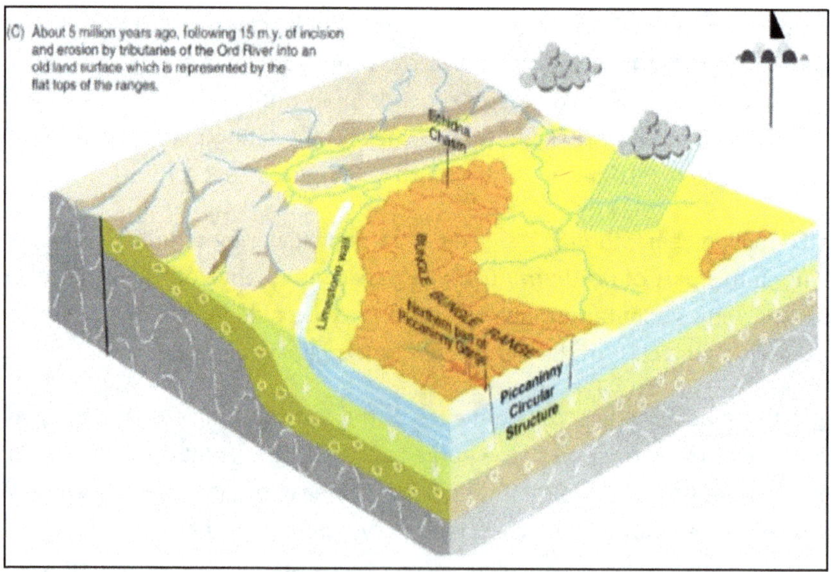

Figure 29. Up to about 5 mya after 15 to 20 my of erosion, the Ord River and its tributaries cut into the old land surface leaving flat topped ridges. Source: Tyler and others 2012.

Such localized erosion over millions of years has allowed flowing water to cut deeply into the Bungle Bungle Range producing the long narrow gorges seen today.

Landforms of the Bungle Bungle Range

The Bungle Bungle Range is a plateau partly bounded by cliffs and cut by numerous gorges which is surrounded by an extensive sand plain. Sandstone beehives characterize the northeastern and southwestern parts of the range — well-developed beehives actually make up less than 20 per cent of the total area of the Bungle Bungle Range. Near Picaninny Creek, they form prominent tiers rising up to the summit plateau. Some beehives are joined together in lines, some form clusters with short side ridges separated by box-like flat-floored valleys, and some occur as isolated groups on the sand plain surrounding the main range. Because of their extent and variety in shape and size, the sandstone beehives of the Bungle Bungle Range are the best examples of this type of landform in Australia and one of the most outstanding examples in the world.

Deep gorges are prominent in the dissected summit area in the northwest. Examples include Echidna Chasm and Mini Palms Gorge. The gorges have sheer sides, extend for distances of up to several kilometers, and are often only a few meters wide. High cliffs of the western escarpment mark the western edge of the Bungle Bungle Range. The cliffs stand up to 200 m high and are cut by several deep gorges. The escarpment dominates the landscape along the Gorge Track and the view from Walanginjdji Lookout (see Figures 22 & 23 for escarpment features).

The weakly dissected plateau of the Bungle Bungle Range rises to a maximum height of about 250 m above the surrounding plain in the northwest (to just over 600 m above sea level) where it consists of smoothly rounded bulbous domes of conglomerate. The plateau surface steadily decreases in height towards the south where it is developed on closely jointed sandstone. The plateau area includes the Piccaninny Circular Structure (site of a meteorite impact?).

Broad valleys up to 500 m wide occur in the southwestern and northwestern parts of the range. The valley of Picaninny Creek which penetrates many kilometers into the center of the range is the longest. The flat to gently undulating sand plains surrounding the Bungle Bungle Range cover about two-thirds of the Purnululu National Park. The plains decrease in height by about 100 m southwards from the base of the range down to the Ord River. Sand dunes on the plains are most prominent at the southern end of the range.

In their size and shape, the sandstone beehives of the Bungle Bungle Range resemble landforms found in many limestone karst areas. However, limestone landforms result from chemical weathering — the calcium carbonate of the limestone is dissolved by water passing over and through the rock whereas the spectacular beehives at Purnululu have been created by the physical removal of grains from the sandstone by mechanical weathering — the action of wind, rain, plants, animals, and particularly flowing water.

The sandstone forming the beehives is extremely friable — sand grains can be easily rubbed off unweathered sandstone, even with one's fingers. This is due to the almost complete absence of cement such as silica, clay, or carbonate between the individual sand grains in the sandstone. Cement was possibly present, but was later dissolved and removed during weathering. Samples examined of the sandstone under very high magnification with a scanning electron- microscope noted that the sand grains in this sample were heavily etched and had quartz overgrowths concluding that the lack of cement was due to the corrosive action of solutions passing between sand grains during weathering. An alternative view is that most of the sandstone has never had much cement. In any case, because it is so friable, the sandstone is particularly susceptible to mechanical erosion. When hit by a hammer, or scraped by a boot, the sandstone disintegrates into a pile of loose sand. Consequently climbing on the beehives in Purnululu National Park is strictly prohibited. In spite of being so friable, the sandstone is able to maintain relatively stable vertical cliffs and steep-sided beehives. This is because the individual grains in the sandstone touch and interlock with one another.

The sites of sandstone beehives in the Bungle Bungle Range are largely controlled by vertical joints in flat-lying sandstone in the southwestern and northeastern parts of the range. In the Picaninny Gorge-Cathedral Gorge area the shapes and positions of the beehives are controlled by two major sets of joints, one trending northwest and the other trending northeast. Many of the joints are lines of weakness for erosion, but some have been impregnated with silica making them more resistant to erosion than adjacent sandstone — these joints stand out as sharp thin ridges or ribs on the beehives. In the northwestern part of the range, in the Echidna Chasm and Mini Palms Gorge area, the rock is mainly conglomerate which is not as closely jointed as the sandstone. No beehives are present and the area is characterized by deep gorges (**Figure 30**).

Figure 30. The entrance to Cathedral Gorge displays exposures of the Bungle Bungle beehives. The left side exposure displays layers of reddish weathered sandstone intermixed with broken fresh sandstone layers that are whitish in color. The broken rock indicates the sandstone is fragile.

The sides of many beehives and cliff faces have hollows formed where sand grains, pebbles, and larger fragments have fallen away during weathering. These hollows grow larger as weathering continues into the friable sandstone, in places resulting in a hole that passes horizontally through a beehive. The sandstone arch may be an example of this process of 'cavernous weathering'. Other examples include the oven-shaped caves in the cliffs above the entrance to Echidna Chasm and on a much smaller-scale, the 'honeycomb' weathering seen along the sandstone walls of Cathedral Gorge (**Figure 31**).

The mystery of the banding on the beehives

A distinctive feature of the sandstone beehives is their alternating dark grey and orange horizontal banding. The bands are parallel to the sedimentary layering (bedding) in the sandstone, and can be traced from beehive to beehive for many kilometers. Each band is up to a few meters thick and consists of many individual sandstone beds. Although the bands appear to be sharply defined from a distance, the contacts between them are seen to be somewhat irregular when closely examined.

Figure 31. The Echidna Chasm exposes the walls of the calcite cemented sandstone. The chasm is actually the end of a fault structure.

The dark grey bands have a protective coating which has been identified as cyanobacteria (formerly known as 'blue-green algae'). No lichen or true algae were present in the samples examined. The cyanobacteria are extremely small (0.01— 0.025 mm) single-celled or colonial organisms that can survive dry periods. The coatings are dull grey during the dry season, but become shiny dark green to almost black after rain. In contrast, the orange bands have protective coatings of iron-oxide cemented sandstone. Both types of coating are only a few millimeters thick, but they play an important role in retarding weathering and erosion of the underlying friable white to pale grey sandstone. Why only some sandstone bands have coatings of cyanobacteria may be attributed to a delicate balance between the climate and slight differences in the physical properties of the sandstone such as its clay content and porosity. The cyanobacteria will grow on the sandstone surfaces that remain moist long enough for their survival not only on the sides of beehives but also in creek beds and on waterfalls. The orange bands, though, are formed on sandstone surfaces which dry out too quickly for the cyanobacteria to grow, and have become coated with iron-oxide instead.

Is the Bungle Bungle Range landscape unique?

Sandstone beehives similar to those of the Bungle Bungle Range are known in the Sahara Desert of North Africa and also in other parts of northern Australia e.g. Mirima (Hidden Valley) National Park near Kununurra, the Keep River National Park, 30 km east of Kununurra, the 'Ruined City' of Arnhem Land, Northern Territory, and the Southesk Tableland, 450 km south of Halls Creek in the Canning Desert.

However, those of the Bungle Bungle Range are unrivalled in terms of their extent, size, variety of shapes, and color banding (**Figure 32**).

The Picaninny Circular Structure — site of a meteorite impact?

The Picaninny Circular Structure in the central part of the Bungle Bungle Range has had a considerable effect on the local landforms and drainage. This elliptical-shaped feature is 7 to 7.5 km across with its long axis aligned northwest-southeast. It has a central elevated area about 4 km across, and up to 50 m higher than an encircling ring depression (**Figure 33**).

Figure 32. Left- Mirima National Park dome shaped sandstone structures similar to the Bungle Bungles. Right – Nigli Gap in Keep River National Park dome structures. Source: TripAdvisor (left), 4WDAUS (right) posted on the internet.

The sandstone in the central area is more fractured, folded, and silicified (impregnated with silica) than that outside. A thin ridge of silicified sandstone marks the southeastern edge of the structure. Some vertical joints and fractures in the surrounding sandstone change direction or terminate at the outer margin of the structure. The sandstone within the circular structure is less dissected than that in other parts of the Bungle Bungle Range, indicating that it may be less friable here than elsewhere.

Figure 33. The suspect Picaninny Crater is thought to be a meteorite impact structure based on its circular structure.

The abrupt change in the course of Picaninny Creek from northeast to northwest upstream from The Elbow can be attributed to the presence of more-resistant rocks in the circular structure.

A favored theory for the origin of the Picaninny Circular Structure is that it represents the deep root of a meteorite impact structure, the upper part of which including the original crater and large blocks of ejected rock has been removed by erosion. According to this theory, the present central elevated area and outer ring depression correspond to the uplifted central peak and the depressed rim characteristic of high-velocity impact craters more than 4 km across. The meteorite is believed to have collided with the Earth sometime between 300 million and 180 million years ago when the sandstone and conglomerate that form the Bungle Bungle Range were buried beneath younger rocks several kilometers thick.

Single Day Tour: Morning

Visit ranger station/information and sales shop (souvenir items only).Take the scenic helicopter flight over Bungle Bungle Range from the helipad at the airstrip over Cathedral Gorge (**Figure 34**).

Figure 34. The inner Cathedral gorge exposes the vertical cathedral structure in the center right which is coated with cyanobacterial coatings (darker black) on top of iron oxide cement (reddish orange). Water cascades down the sides of the cathedral eroding away the surrounding rock leaving the center part exposed. The block on the left appears to be undermined by higher water levels. The pool in the foreground represents the plunge pool from water falls cascading down into the cave from the foreground roof and background wall. Source: Rob Cox posted on the internet.

Afternoon

Visit the Echidna Chasm (for a mid-day sun experience) and/or Froghole Gorge.

Leave Purnululu National Park on Spring Creek Track to the Great Northern Highway (2.5 hours).

Day one: Morning

Visit ranger station/information and sales shop. Take a scenic helicopter flight over the Bungle Bungle Range from the helipad at the airstrip over Cathedral Gorge.

Day one: Afternoon

Echidna Chasm (for a mid-day sun visit). Mini Palms Gorge, Froghole Gorge, or Walanginjdji Lookout (for a sunset visit) (**Figure 35**).

Figure 35. View looking into Mini Palms Gorge. Source: Redbubble posted on the internet.

Day two: All Day

Picaninny Gorge (walk to the Black Waterfall with a 16k return). (**Figure 36**).

Leave Purnululu National Park on Spring Creek Track to the Great Northern Highway (2.5 hours).

Figure 36. Picaninny Gorge. Source: going feral one day at a time posted on the internet.

Three-day panorama tour

Picaninny Gorge — camping overnight in the gorge (exploring side gorges upstream from The Fingers region with a 30+ km return trip).

Leave Purnululu National Park on Spring Creek Track to the Great Northern Highway (2.5 hours)

Unlimited tour of Purnululu National Park

Visit ranger station/information and sales shop. Take the scenic helicopter flight over Bungle Bungle Range from the helipad at the airstrip. Following the helicopter tour, visit the following sites:

 Domes Trail Cathedral Gorge (inner gorge walking tour)

Echidna Chasm (for a mid-day sun visit)

Froghole Gorge

Mini Palms Gorge

Walanginjdji Lookout (for a sunset visit)

Picaninny Gorge — camping overnight in gorge (exploring side gorges upstream from The Fingers region — 30+ km return trip).

Walking trails in order of increasing distance (return journey included)

Walanginjdji Lookout 0.5 km

Domes Trail 1 km

Froghole Gorge 1 km

Echidna Chasm 2 km

Cathedral Gorge 3 km

Mini Palms Gorge 5 km

Picaninny Gorge 14 km (up to The Elbow)

Piccaninny Gorge 30+ km

Walking trails in order of increasing time (return journey included)

Walanginjdji Lookout - 30 minutes

Domes Trail - 45 minutes

Cathedral Gorge - 1 to 2 hours

Echidna Chasm - 1 to 2 hours

Froghole Gorge - 1 to 2 hours

Mini Palms Gorge - 3 hours

Picaninny Gorge (up to The Elbow) - 1 day

Picaninny Gorge (above The Fingers) - 2+ days

Walking trails in order of increasing difficulty (scale of difficulty: 1 = easy, 5 = difficult)

Domes Trail 1

Walanginjdji Lookout 1

Cathedral Gorge 2

Echidna Chasm 3

Picaninny Gorge 3 (up to The Elbow)

Froghole Gorge 4

Mini Palms Gorge 4

Picaninny Gorge 5 (above The Elbow)

Echidna Chasm

Features: Spectacular long narrow chasm; *Livistona* palms; towering cliffs with oven shaped caves near entrance; large fallen blocks of conglomerate; striking lighting effects and rich earthy colors for photography. Location: Northwestern side of the Bungle Bungle Range about 20 km along the track from the ranger station. Allow for a distance of 2 km for the return; allow 1 to 2 hours. Facilities: Toilet at Echidna Chasm carpark. Grade: Easy to moderate; most difficult near the end where the gorge is partly blocked by boulders (**Figure 37**).

Figure 37. Trail leading into Echidna Chasm from the eastern most carpark.

Echidna Chasm is one of the most popular walking trails in the Bungle Bungle Range. It takes its name from an echidna found nearby several years ago. The parking area at the start of the walking trail is in front of a high cliff of pebbly sandstone. The sandstone surface is generally iron-stained (weathered) to yellowish, reddish, or brownish colors. Unstained clean sandstone almost pure white is revealed at the base of the cliff where a recent rock fall has taken place (point of interest 1). The walking trail from the parking area skirts to the left (eastwards) around the base of the cliff to a pebbly creek bed which it follows upstream around large fallen blocks of conglomerate (point of interest 2) to the entrance of Echidna Chasm. Many oven-like caves formed by cavernous weathering occur in the towering cliffs on either side of the chasm entrance.

Majestic tall and slender *Livistona* palms grow in the bed of the creek and along its banks, on steep debris slopes to either side, and also on narrow ledges and in crevices high up the cliff faces.

On entering the chasm — a vertical slot over 180 m high, but nowhere more than a few meters wide — the trail continues along the creek bed for several hundred meters, through the upper part of the chasm (point of interest 3) before ending abruptly at a vertical rock face.

To reach the end, it is necessary to clamber over fallen blocks of conglomerate and beneath similar blocks that have become wedged between the sides of the chasm. Looking up at the end of the chasm to the narrow opening far above, palm fronds can be seen glinting in the sunlight. Throughout its length, the chasm is in deep shade except around late morning when the sun is vertically overhead — the best time for photography.

Froghole Gorge

Features: *Livistona* palms; high cliffs; fallen blocks of conglomerate; lightning strike on ridge crest; sheer waterfall with plunge pool inhabited by frogs; ripple- marked sandstone and small faults in cliff face. It is located on the Northwestern side of the Bungle Bungle Range about 19 km along the track from the ranger station. Distance along the trail is 1 km including the return; allow 1 to 2 hours. Facilities: Toilet and shelters at Froghole Gorge carpark. Grade: Moderate to difficult walk with a steep slope and large fallen blocks of conglomerate to be negotiated near the end of the gorge.

Froghole Gorge is a shorter but more demanding walk than that into the nearby Echidna Chasm. The walking trail leaves the parking area at the entrance to Froghole Gorge and follows a pebbly creek bed upstream to the head of this relatively open gorge. Several diversions are needed to pass large fallen blocks of conglomerate (point of interest 1). Unweathered white sandstone is exposed in a large undercut near the base of the cliff on the left-hand (northern) side of the trail (**Figure 38**).

A patch of white loose sand on the ridge to the right on the western side of the gorge just below a jagged sandstone knob was formed when a bolt of lightning struck the knob, probably in 1991 (looking southwest from point of interest 2). Originally flat-lying, the sandstone and conglomerate here have been tilted by movement along faults to the northwest. The trail ends at the base of a vertical cliff with a chute-like waterfall which is only active after heavy rain (point of interest 3).

53

A semi-permanent pool at the base of the waterfall is home for several species of small frogs — hence the name of the gorge. Ancient ripple marks preserved in a sandstone ledge directly above the pool were formed by water flowing over sand 360 million years ago. To the left of the waterfall near ground level a small fault with a displacement of about 1 m can be seen in the cliff face.

Figure 38. From the Froghole carpark, trail leads into the Froghole Gorge.

Mini Palms Gorge

Features: Pebbly creek beds; high cliffs; fallen blocks of conglomerate; spectacular amphitheater with stunted *Livistona* palms; cave to be explored by torch light. The location is on the Northwestern side of the Bungle Bungle Range about 19 km along the track from the ranger station. The trail commences from the Froghole Gorge carpark.

Distance: 5 km return; allow 3 hours. Facilities: Toilet and shelters at Froghole Gorge carpark. Grade: Easy at first along gravel creek beds but becoming difficult as steep slopes and large blocks of fallen rock have to be negotiated.

The walking trail to Mini Palms Gorge leaves the carpark at the entrance to Froghole Gorge proceeding to the right (westwards) over sand and spinifex along the base of the main range to join the creek system that comes from Mini Palms Gorge and nearby gorges. This first part of the trail follows the boundary (a fault concealed by sand) between the Devonian sandstone and conglomerate which form the Bungle Bungle Range to the left, and older rocks that are Cambrian in age mostly hidden beneath sand to the right. The cliff-forming Devonian sandstone nearest the fault is cut by many quartz-filled fractures which developed when movements took place along the fault many millions of years ago. The Cambrian rocks include grey limestone forming a low ridge on the right-hand (northern) side of the walking trail (point of interest 1). This is the same limestone that forms the prominent narrow wall along the Gorge Track north and south of the Three-Ways Junction (**Figure 39**).

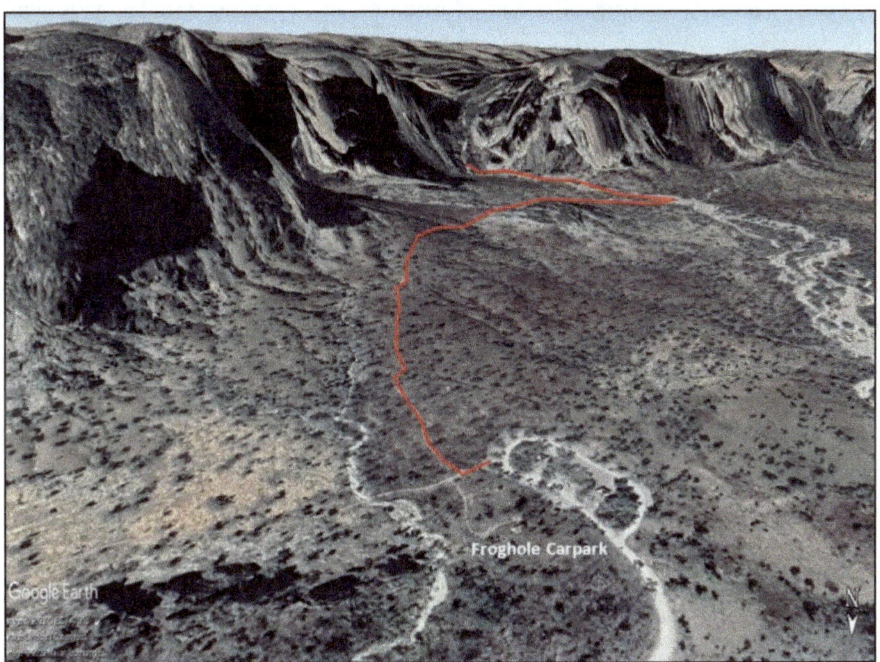

Figure 39. Trail leading into Mini Palms Gorge from the Froghole carpark.

The next part of the trail follows pebbly creek beds upstream (southwestward) through a broad gap between the main range and an outlying sandstone ridge (point of interest 2). The sandstone knob struck by lightning on the west side of Froghole Gorge trail can be seen on the ridge crest to the east. After crossing a relatively large open area (point of interest 3), walkers have to squeeze between conglomerate boulders to enter a smaller open area (point of interest 4) where several creeks come together.

The trail continues up the main creek (southwards) before climbing a steep rock-fall that blocks the entrance to Mini Palms Gorge.

From the top of this rock-fall there is an awe-inspiring view into Mini Palms Gorge: a high narrow amphitheater with stunted *Livistona* palms growing on the flat sandy floor (point of interest 5). After scrambling down onto the floor and crossing to the far end of the amphitheater, taking care not to cause any damage to the stunted palms, there is a narrow cave to be explored by torchlight. A gentle breeze in the cave comes from the hidden opening of an inaccessible gorge upstream. The smooth sides of the amphitheater and large vertical cracks in the rock face near the mouth of the cave are joints — planes along which the rocks have been preferentially eroded.

The stunted *Livistona* palms on the floor of Mini Palms Gorge gradually increase in height both towards the front of the amphitheater and inwards from the side walls — towards where the light is brightest. This observation indicates that poor light is probably the main reason for the stunting, although short-lived flooding during the wet season, when the floor becomes covered by more than a meter deep of water, may be a contributing factor.

Kungkalanay Lookout

Features: Panoramic views to the north, east, south, and west; particularly photogenic in the late afternoon. Location: Western side of the Bungle Bungle Range about 1.0 to 1.5 km along the Gorge Track from the ranger station. Distance: 500m return; allow 30 minutes. Facilities: None. Grade: Easy gradual climb.

Situated on the crest of a ridge, the Kungkalanay Lookout provides photogenic panoramic views in all directions. To the east, the majestic cliffs of the western escarpment of the Bungle Bungle Range form the backdrop to low spinifex- covered ridges of basalt in the foreground, a low wall-like ridge of grey Cambrian limestone in the middle distance and a higher ridge of younger dark reddish Cambrian sandstone farther back. At sunset, the high cliffs and a prominent pyramidal hill of sandstone at the northern end become ablaze with brilliant red and yellow colors in marked contrast to the subdued colors of the shaded foreground (**Figure 40**).

The Osmond Range is visible on the skyline to the north and northwest (**Figure 41**). To the west are the rolling hills and steep-sided ridges through which the Spring Creek Track winds its way from the Great Northern Highway to the Purnululu National Park (**Figure 42**).

The Dixon Range made of the same Devonian sandstone that forms the Bungle Bungle Range, and the valley of the Ord River can be seen in the far distance to the south beyond the broad sand plain (**Figure 43**).

Figure 40. Facing east, the Bungle Bungle Range western scarp is visible.

Figure 41. The view to the north consists of the sand plain to the left, small hills consisting of the Osmond Hills.

Figure 42. View facing west of the rugged landscape in the background which the Spring Creek Track passes through.

Figure 43. View facing south of the sand plain and western Bungle Bungle Range on the left.

Southwestern part of the Bungle Bungle Range

Erosion of the flat-lying Devonian sandstone in the northeastern and southwestern parts of the Bungle Bungle Range has left a complex of horizontally banded conical towers with rounded tops — the sandstone beehives for which Purnululu is justly famous. Some of the sandstone bands contain scattered pebbles, and there are thin lenses of conglomerate in places, but no thick conglomerates like those in the northwest. The alternating orange to red-brown and grey to black banding is a surface feature (see The mystery of the banding on the beehives). The orange to red-brown bands are coated by iron oxide whereas the grey to black bands belong to protective skins of cyanobacteria. Grey nests of termites are a feature of many beehives. They include tall and slender free-standing mounds, squat masses, and irregular trails plastered onto the sides of the beehives. The material forming the nests includes grains of sand derived from the beehives, and organic material such as spinifex and probably also cyanobacteria.

The main watercourse in this part of the range is Picaninny Creek which is fed by several smaller creeks including that from Cathedral Gorge. The locations of these creeks are largely controlled by major joints in the sandstone.

Some joints are readily eroded planes of weakness, but others have been impregnated with silica and are relatively resistant to erosion.

The southwestern gorges support a generally sparse vegetation that includes spinifex up to 2 m high, small shrubs, rock figs, and small trees of *Eucalyptus* species. *Livistona* palms grow in Picaninny Gorge.

Picaninny Gorge

Features: Sandstone beehives of all shapes and sizes; high cliffs; termite nests; creek beds with fluted pavements, potholes, pebble accumulations, seasonal waterholes and waterfalls. Location: Southwestern side of the Bungle Bungle Range about 25 km along the track from the ranger station. Distance: 14 km return to The Elbow; 30+ km return from the many side gorges above The Fingers. Allow several days, camping overnight, to explore the side gorges above The Elbow. Register with the ranger before commencing overnight walks to obtain guidelines for camping. Facilities: Toilet and shelters at Picaninny Gorge carpark. Grade: Generally easy to The Elbow, as the trail is mainly along firm creek pavements; moderate to difficult upstream from The Elbow where the gorge narrows and becomes partly blocked by fallen rocks, and loose pebbles and cobbles cover the creek bed. Deep waterholes occur in narrow gorges upstream from The Fingers. Take sufficient drinking water, sun protection, and suitable footwear (**Figure 44**).

The Picaninny Gorge trail is a challenging walk which explores the longest gorge in the Bungle Bungle Range. The gorge derives its name from the beehives which according to Aboriginal mythology are the babies of the mother Bungle Bungle Range. The carpark at the beginning of the trail gives a good view of sandstone beehives rising in tiers northwards towards the high summit area of the main range. The walking trail initially follows part of the Domes Trail passing close to a beehive with a hole through it (see point of interest 1 for the Domes Trail), and then continues upstream along the bed of Picaninny Creek passing close to many beehives and side gorges. The first 7 km of the trail runs generally northeastwards with many abrupt changes in direction as far as a major right-angle bend which is The Elbow. Here, Picaninny Creek swings sharply northwest along the southwestern side of the Picaninny Circular Structure continuing up to The Fingers area where several major side gorges join the main gorge.

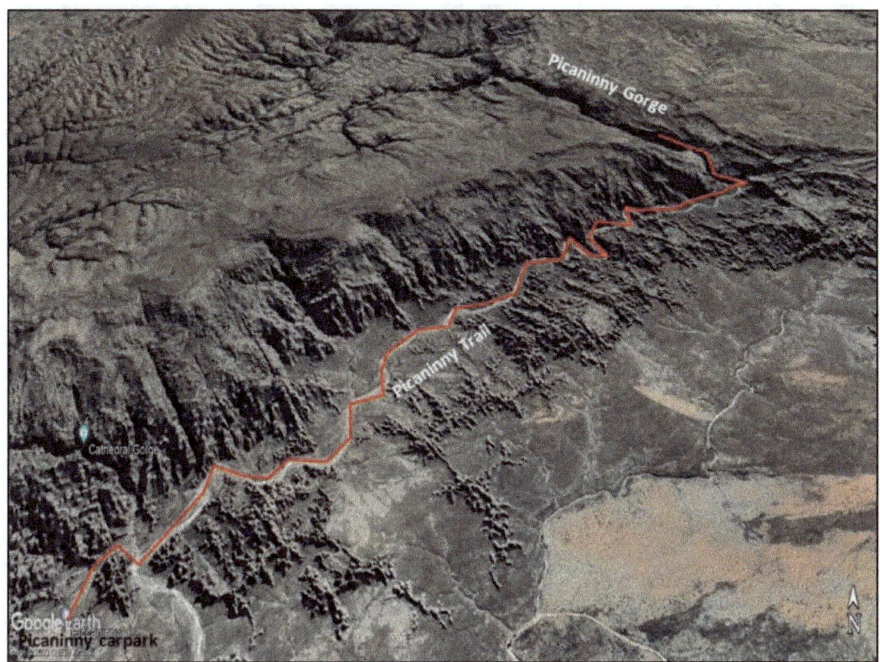

Figure 44. The Picaninny Trail leads from the Picaninny carpark in the southwest corner of the image to the Picatininny Gorge in the upper right of the image. The trail follows the Picatininny Creek bed.

The trail up to The Elbow is relatively easy going, as it is mainly on solid sandstone bedrock, although there are some patches of loose cobbles and pebbles. Fluted creek beds and deep potholes (points of interest I, 2, and 4) are characteristic features. The long parallel flutes in the creek bed have been formed by pebbles and cobbles rolling downstream in swiftly flowing floodwater. The circular potholes have been worn by tumbling pebbles trapped in eddies.

Some of the largest potholes are located at the downstream ends of straight creek reaches where the creek takes an abrupt change in direction, or below sudden drops in the level of the creek bed. Seasonal and semi-permanent pools of water occur on some sharp bends and below waterfalls especially where shaded from the sun. All walkers are urged to avoid polluting these and all other waterholes in the national park.

Well-preserved sedimentary features in the flat-lying Devonian sandstone exposed in the bed of Picaninny Creek indicate that this rock was deposited by ancient rivers. These features include cross-bedding, desiccation cracks, and current ripple marks, all of which can be seen at point of interest 3 (**Figure 45**).

Figure 45. Devonian sandstone exposed in the Picaninny Creek at the beginning of the trail east of the carpark.

Looking downstream from just above The Elbow (point of interest 5) there is a line of three relatively small beehives that resemble in profile the famous Three Sisters landmark in the Blue Mountains of New South Wales.

Upstream from The Elbow, the relatively broad valley of Picaninny Creek narrows to become Picaninny Gorge — a fairly straight gorge between towering cliffs with large fallen blocks of pebbly sandstone and groves of *Livistona* palms. Walking along the first part of the gorge is hampered by loose pebbles and cobbles underfoot, and long stretches are exposed to the hot afternoon sun.

However, there are some idyllic shady spots like the permanent waterhole below Black Waterfall (point of interest 6) on the right-hand (northeastern) side of the trail about 1 km upstream from The Elbow. Farther upstream, at The Fingers (point of interest 7), the main gorge splits into several sheltered side gorges containing large waterholes and lush vegetation.

Features: Short walk; sandstone beehives; creek pavements with potholes; towering cliffs; slipped sandstone blocks; honeycomb weathering; waterfalls (generally dry); large amphitheater with reflective pool of water and white sandy beach. Location: Southwestern side of the Bungle Bungle Range about 25 km along the track from the ranger station. The trail commences from the Picaninny Gorge carpark. Distance: 3 km return from the Picaninny Gorge carpark; allow 1 to 2 hours. Facilities: Toilet and shelters at Picaninny Gorge carpark. Grade: Easy to moderate; some short steep slopes and narrow ledges around waterholes (**Figure 46**).

Figure 46. The Cathedral Gorge Trail is accessed from the Picaninny carpark.

The walking trail to Cathedral Gorge branches left from the Picaninny Gorge trail about 600 m from the Picaninny Gorge carpark. It follows a creek bed, and involves some clambering over rock ledges and around large potholes some of which may contain water.

Polygonal jointing or 'crazy paving' in creek beds as found near the entrance to the main gorge (point of interest 1) is a result of present-day alternating expansion and contraction of the sandstone bedrock caused by the regular seasonal wetting and drying.

Another notable feature to look for nearby on the right-hand (northeastern) side of the trail is the extensive and complex root system of a small fig tree on the side of a low rock wall. Farther along the trail in the lower parts of the sandstone walls in the narrower part of the gorge there are many fine examples of honeycomb weathering (see How the beehives formed).

The role of joints in the erosion of the gorge is well displayed upstream. On the left-hand (southern) side of the gorge, large rectangular slabs of sandstone (the 'fallen warriors') bounded by joints have slipped several meters from the rock face above (point of interest 2). Some of the joints can be traced from here for about 100 m to the right-hand side of the large Cathedral Gorge amphitheater — the Cathedral — at the head of the gorge sandstone fallen from the ceiling lies below. A large pool of water on the floor shaded from the sun and surrounded by white sand is filled during the wet season via a waterfall. The water in the pool provides stunning photogenic reflections of the soaring sandstone beehives 'guarding' the entrance to the amphitheater. The acoustics of the Cathedral are such that many self-proclaimed singers have tried to impress others with their renditions.

Domes Trail

Features: A short easy walk in the Bungle Bungle Range; clusters of sandstone beehives including one with a hole through it; creek bed with polygonal jointing. Location: Southwestern side of the Bungle Bungle Range about 25 km along the track from the ranger station. Includes first part of the Cathedral Gorge and Picaninny Gorge trails. The trail commences from the Picaninny Gorge carpark. Distance: 1 km circuit trail; allow 45 minutes. Facilities: Toilet and shelters at Picaninny Gorge carpark. Grade: Easy; mostly flat and partly along firm creek pavements (**Figure 47**).

This short circuit enables visitors to walk among the beehives that arise abruptly from the spinifex-covered sand plain near the Picaninny Gorge carpark. The trail is on level ground, and there are no steep slopes or loose gravelly creek beds to deter the less energetic walker. A notable feature is a beehive with a large hole eroded horizontally through it (point of interest 1). A short side trail in the northwest leads into a typical blind gorge flanked by towering beehives (**Figure 48**).

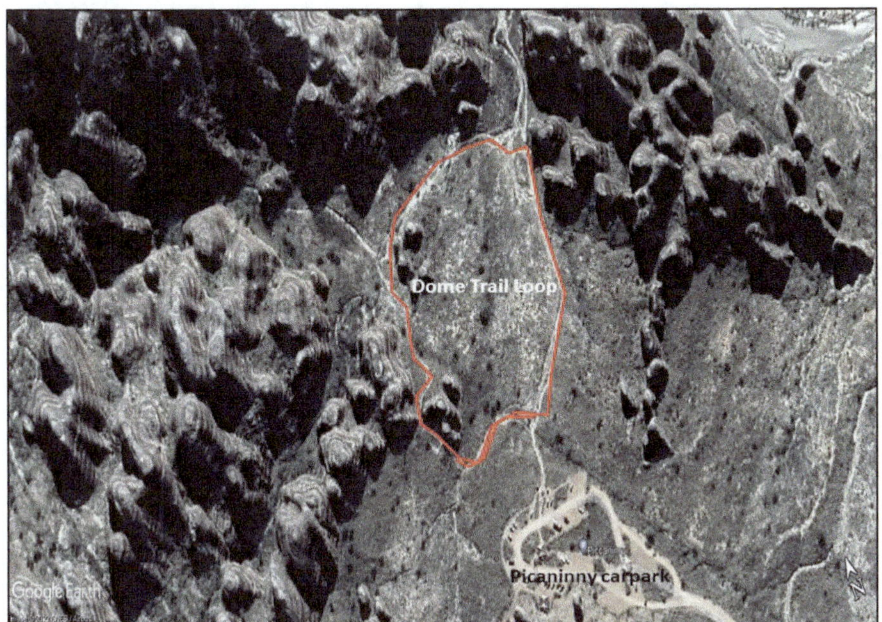

Figure 47. The Dome trail is close to the Picaninny carpark circling on the north side in front of the Bungle Bungles range.

Figure 48. In the southwest corner of the dome Trail where the trail passes through a narrow beehive position on both sides of the trail, the stratigraphic sequence of the beehive can be observed on either side of the trail. This image was taken from the west side of the trail.

References

Hoatson, D., Blake, D., Mory, A., Tyler, I., Piita, M., Kamprad, J., Oswald-Jacobs, I.O., 1997. Bungle Bungle Range Purnululu National Park, East Kimberley, Western Australia: a guide to the rocks, landforms, plants, animals, and human impact. Geological Survey of Western Australia posted on the internet including references contained within.

Tyler, I.M., Hacking, R.M., Haines, P.W. 2012. Geological evolution of the Kimberley region of Western Australia. Geological Survey of Western Australia including references contained within.

Aussie Gold:
The geology behind lode and alluvial
gold deposits and associated minerals
in Western Australia
William A. Szary, M.S.
Copyright 2021, Earth2Energy
Educational Publishing.
All rights reserved.

Aussie Gold: Victoria Province
The geology behind lode and alluvial
gold deposits and associated minerals
William A. Szary, M.S.
Copyright 2021, Earth Energy
Educational Publishing.
All rights reserved.

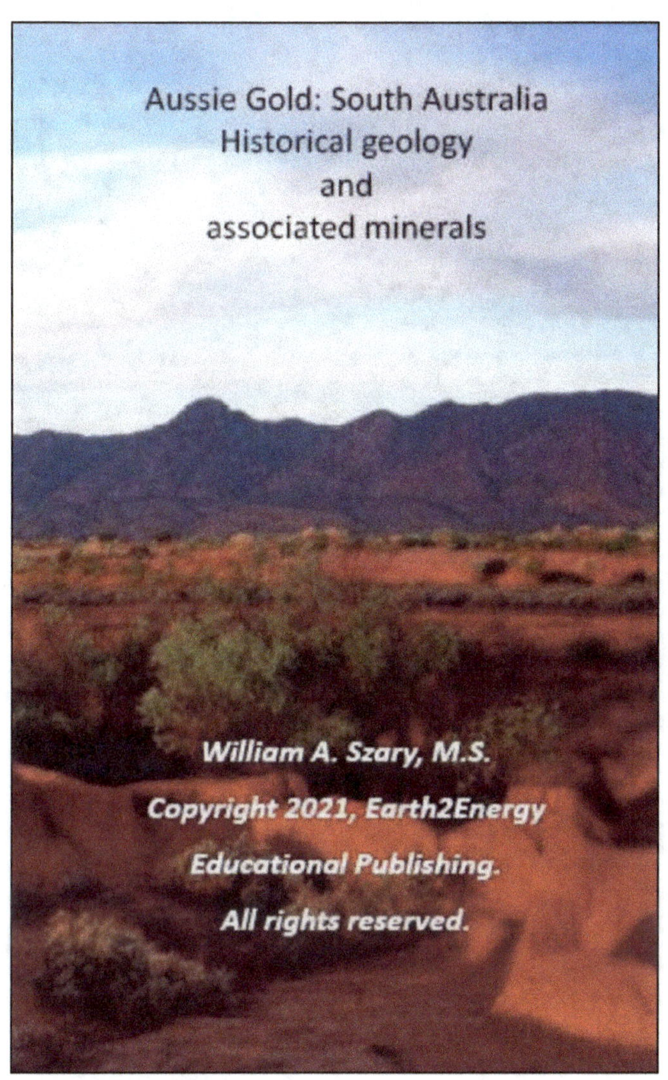
Aussie Gold: South Australia
Historical geology
and
associated minerals

William A. Szary, M.S.
Copyright 2021, Earth2Energy
Educational Publishing.
All rights reserved.

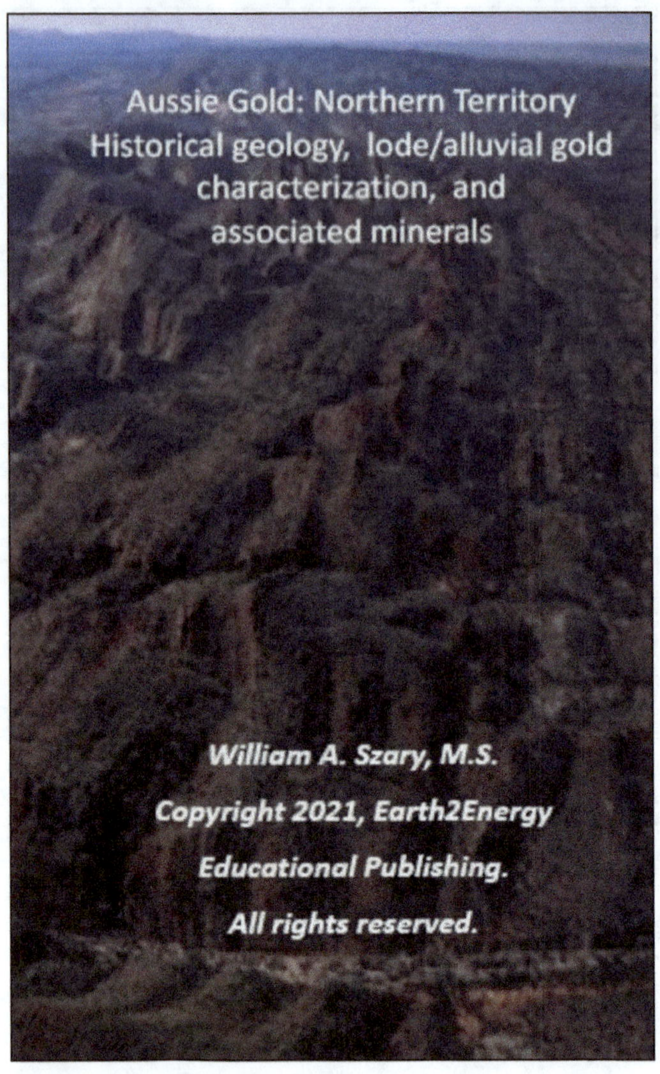

Aussie Gold: Northern Territory
Historical geology, lode/alluvial gold
characterization, and
associated minerals

William A. Szary, M.S.

Copyright 2021, Earth2Energy

Educational Publishing.

All rights reserved.

Aussie Gold: New South Wales
Lode and Alluvial Gold, Sapphire, and
Diamonds

William A. Szary, M.S.

Copyright 2022, Earth2Energy

Educational Publishing.

All rights reserved.

Aussie Gold: Queensland
Geological Framework
and
Mineralization

William A. Szary, M.S.

Copyright 2022, Earth2Energy

Educational Publishing.

All rights reserved.